D0969014

American Hazardscapes

The Regionalization of Hazards and Disasters

Susan L. Cutter, *Editor*

JOSEPH HENRY PRESS
Washington, D.C.

JOSEPH HENRY PRESS • 2101 Constitution Avenue, N.W. • Washington, D.C. 20418

The Joseph Henry Press, an imprint of the National Academy Press, was created with the goal of making books on science, technology, and health more widely available to professionals and the public. Joseph Henry was one of the founders of the National Academy of Sciences and a leader in early American science.

Library of Congress Cataloging-in-Publication Data

American hazardscapes : the regionalization of hazards and disasters / edited by Susan L. Cutter.
p. cm.—(Natural hazards and disasters)
Includes bibliographical references.
ISBN 0-309-07443-6
1. Natural disasters—United States. 2. Disasters—United States.
3. Environmental risk assessment—United States. 4. Emergency management—United States. I. Cutter, Susan L. II. Title. III. Series.
GB5010 .A43 2001
363.34′0973—dc21

2001003606

Printed in the United States of America.

Preface

The environment poses certain threats to our well-being, yet many of us continue to choose to live in risky or dangerous places such as barrier islands, floodplains, or along known earthquake faults. There is a host of other, more "hidden" hazards related to environmental contamination that equally pose serious threats to our health and well-being. Both contribute to America's hazardscape—the landscape of hazards that we find locally, regionally, and nationally.

Where are these places and what types of hazards are found there? This ***where*** question requires a distinctive perspective, that of a geographer, to help us understand the distribution of these hazards and the regional patterns they produce. Further, the geographer can also assist in mapping and rendering cartographic representations of hazardous environments and communicate this hazard information to the public and policy makers. The ***what*** question requires a hazards researcher's skill in comprehending the interaction between society, the natural environment, and the built environment and how we respond and adjust to such interactions over time. The hazards researcher helps us to realize the short-term adjustments and longer-term adaptations that individuals and society must make to create more "disaster-resistant" communities and how these adjustments and adaptations may require regional specificity, depending on the nature of the threat.

This book, the fifth in the Natural Hazards and Disasters series published by the Joseph Henry Press, is derived from these two perspectives—the geographer's and the social science hazards researcher's. Our goal is to illustrate what we know about the patterns of hazard events and losses during the previous three decades and how we improve our understanding of the events themselves as well as their impact on society. The book begins with an overview of the development of hazards and risk research and its use in reducing losses from environmental threats. A review of the methods for and innovations in hazard and vulnerability assessment follows in Chapter 2. A brief history of mapping and the spatial analysis of hazards and risks are presented in Chapter 3. To provide an understanding of the temporal and geographic nature of hazard events and losses, we next describe the availability, quality, and usefulness of national data sets on hazard events and losses (Chapter 4). A retrospective look at trends in hazard events and losses over the past 24 years is found in Chapter 5. Chapter 6 examines the geographic variations in hazard events and losses at the state level, thereby developing a regional ecology of disaster-prone or disaster-resistant states. The last chapter offers some thoughts on what local, state, and federal managers need to do to meet the challenges of reducing hazard losses during this century.

This book would not have been possible without the support and guidance of Dennis Mileti who invited us to participate in the Second Assessment Project (funded by the National Science Foundation (NSF), CMS #9312647). We were the "data mavens" for the project, although we often felt like "chefs" trying to figure out what to do with the piece of Swiss cheese we were given (e.g., making sense of a data set full of holes). We also would like to acknowledge the separate support provided by the NSF (CMS #9905352), which enabled us to include updated (1994-1998) data for the book.

There are many people to thank who helped see this book to completion: Michael Scott (Salisbury State University, Salisbury, Maryland); and Betsey Forrest, Jo Darlington, and Mary Fran Myers (Natural Hazards Center, University of Colorado). Stephen Mautner at the Joseph Henry Press has been a supportive (and nonintrusive) editor for which we are most appreciative. The manuscript was significantly improved by the thoughtful comments of reviewers: Dennis Mileti (University of Colorado), Stu Nishenko (FEMA), Michael Helfert (South Carolina State Climatologist), and Gilbert F. White (University of Colorado). We thank you for helping us improve the text and, more importantly, the message of this book. Finally, we owe a debt of thanks to the University of South Carolina Department of Geography and its Hazards Research Lab, especially those research assistants who helped support the database construction and manuscript preparation for the book: Melanie Baker, Patrice Burns, Paul Putnam, Jeff Vincent, Jennifer DeReuss, Jamie Mitchem, and Wilson Brown. Although each author was tasked with developing certain chapters, the book is a true collaborative effort in which we speak with one voice. We therefore take responsibility for any errors of omission, commission, or interpretation. We were honored to be part of the Second Assessment and to be able to contribute in our own special way.

Susan Cutter
Columbia, South Carolina

Contributors

Susan L. Cutter is a Carolina Distinguished Professor in the Department of Geography at the University of South Carolina as well as the Director of its Hazards Research Lab. She is the founding editor of the journal *Environmental Hazards* (Elsevier). Cutter has a long-standing interest in the geographic dimensions of environmental risks and hazards, has published extensively on the topic, and serves on numerous national advisory committees in the field. She was elected a Fellow of the American Association for the Advancement of Science (AAAS) in 1999 for her contributions and served as President of the Association of American Geographers (2000-2001).

Deborah S. K. Thomas is Assistant Professor of Geography at the University of Colorado, Denver. She received her B.A. from the University of Kansas, M.A. from the University of North Carolina, Charlotte, and Ph.D. from the University of South Carolina. She has written on the use of geographic information systems (GIS) in hazards research and applications, children's health and risk, and sustainable environmental planning.

Jerry T. Mitchell is Assistant Professor of Geography in the Department of Geography and Geosciences at Bloomsburg University. He received his B.S. and M.A. from Towson State University and his Ph.D. from the University of South Carolina. He has previously written on the social and cultural dimensions of hazards, including articles on environmental justice, the use of GIS in vulnerability assessments, and the role of religious belief in hazard perception.

Arleen A. Hill is a Ph.D. candidate in the Department of Geography at the University of South Carolina. She received her B.S. from Mary Washington College and her M.S. from the University of South Carolina. She is interested in risk communication and the role of short- and long-term forecasts in prompting evacuation behavior.

Michael E. Hodgson is Associate Professor of Geography at the University of South Carolina and Co-director of the Center for GIS and Remote Sensing. His interests are in the use of geographic techniques (remote sensing and GIS) in hazards analysis and management and ecological modeling.

Contents

American Hazardscapes

CHAPTER 1

The Changing Nature of Risks and Hazards

Susan L. Cutter

During the past three decades, we have seen escalation in the damages caused by natural hazards (van der Wink et al. 1998). It seems as though every year, Americans are recovering from one natural disaster after another—earthquakes (Loma Prieta in 1989, Northridge in 1994), hurricanes (Hugo in 1989, Andrew in 1992, Floyd in 1999), fires (Oakland-Berkeley in 1991), and flooding (Mississippi River in 1993, North Dakota in 1997) to name but a few. Disaster movies, a mainstream genre in Hollywood, add to the perception that disasters are happening more often. The box office success of *The Grapes of Wrath* and the more recent *Twister*, *Dante's Peak*, *Volcano*, and *Armageddon* coupled with a host of made-for-television movies on disaster themes (volcanic eruptions, tornadoes, asteroids), capture our imaginations and lead many people to conclude that disasters are common occurrences, certainly more prevalent today than in the past.

The recent occurrence of many of these events and the hype surrounding fictional disaster movies also leads to a perception that many of these hazards are concentrated in one region or another. Certainly, California immediately comes to mind in both fictional and nonfictional portrayals of a disaster-prone area. The unique marketing gimmick for the movie *Volcano* touts "the coast is toast" when referring to Los Angeles in posters and video trailers. California's image as "the disaster state" is assured in the minds of the mass media as well as most Americans. However, California is not alone. Florida and Kansas (the "Wizard of Oz" effect) are other places that people think are quite disaster-prone. Are these regional stereotypes reflective of the actual occurrence of hazard events? Are we exposed to more hazards now than in the past or are the economic losses from hazards simply escalating and thus capturing our attention?

This book explores both questions through an examination of the temporal and spatial trends in hazard events and losses during the past three decades. We hope to provide some understanding of the changing nature of hazards and hazards assessment, the technological innovations that have improved our ability to display and analyze hazards data, and how these innovations in understanding and technology can reduce local vulnerabilities to hazards through improved mitigation. Our purpose is twofold. First, we want to illustrate the geographic dimensions of hazards—where they occur, why they occur where they do, who is and which places are most vulnerable, and what can be done to reduce local vulnerability. Second, we want to demonstrate the necessity of the hazards (and vulnerability) assessment process and link it to longer-term mitigation and hazards reduction at the local level. Only after society begins to think seriously about the environmental context within which people live and work can we make informed choices regarding the level of hazardousness individuals and communities are willing to bear and at what cost to themselves and the nation.

HAZARD, RISK, AND DISASTER

Within the broad community of hazards researchers and practitioners, hazard, risk, and disaster are terms that are used interchangeably, although they do have different meanings (Cutter 1993, 1994, Kunreuther and Slovic 1996, Quarantelli 1998, Mileti 1999). A *hazard*, the broadest term, is a threat to people and the things they value. Hazards have a potentiality to them (they could happen), but they also include the actual impact of an event on people or places. Hazards arise from the interaction between social, technological, and natural systems. They are often described by their origin—for example, natural hazards (earthquakes) and technological hazards (chemical accidents)— although this classification is losing favor among the research community because many hazards have more complex origins. For example, in many parts of the world, deforestation has resulted in increased runoff, which then leads to catastrophic downstream flooding. Is this a natural or a socially induced hazard? Or consider the use of technology to control nature, such as dams and levees. The levees may hold during normal-rainfall years, but they could fail during abnormally wet years. Is a wet-year levee break and the subsequent flooding that follows a technological, natural, or environmental hazard? As can be seen, hazards are partially a product of society and thus it is impossible to understand hazards without also examining the context (social, political, historic, environmental) within which hazards occur.

Risk is the probability of an event occurring, or the likelihood of a hazard happening (Presidential/Congressional Commission on Risk Assessment and Risk Management 1997). Risk emphasizes the estimation and quantification of probability in order to determine appropriate levels of safety or the acceptability of a technology or course of action. Risk is a component of hazard.

Generally speaking, a *disaster* is a singular event that results in widespread losses to people, infrastructure, or the environment. Disasters originate from many sources, just as hazards do (natural systems, social systems, technology failures). Although there are many perspectives on what constitutes a disaster (Quarantelli 1998), we will stick to the simple definition presented earlier.

As suggested elsewhere (Kates 1978, Whyte and Burton 1980, Krimsky and Golding 1992, Kasperson et al. 1995, Hewitt 1997, Hamilton and Viscusi 1999), the distinction between hazard, risk, and disaster is important because it illustrates the diversity of perspectives on how we recognize and assess environmental threats (risks), what we do about them (hazards), and how we respond to them after they occur (disasters). The emphasis on hazard, risk, and disaster is also reflective of different disciplinary orientations of researchers and practitioners. Historically, the health sciences, psychology, economics, and engineering were concerned about risks—their quantification, mathematical attributes, and use in decision making. Geographers and geologists were primarily interested in hazards, whereas sociologists captured disasters as their intellectual domain. However, as the nature of hazards, risks, and disasters became more complex and intertwined and the field of hazards research and management more integrated, these distinctions became blurred as did the differentiation between origins as "natural," "technological," or "environmental."

EVOLVING THEORIES AND CONCEPTS

In the first assessment of natural hazards research, White and Haas (1975) devoted an entire chapter to the social acceptance and tolerance of risks and hazards. In the second assessment, Mileti (1999) calls for an entirely new philosophical approach in dealing with hazards and reducing losses from disasters. This new perspective emphasizes (1) the interactions among social and natural systems, and the built environment; (2) the notion that hazards and disasters are acts of people, not acts of divine intervention; and (3) that unsustainable environmental practices increase vulnerability to hazards and disasters and thus hinder movement toward sustainability.

White and Haas called for improved research on human adjustments to hazards and dissemination of that research to local and state officials. Mileti says that the shift toward a sustainable approach to hazard mitigation and reduction will require a number of important steps. Among them are (1) conducting a national hazards and risk assessment; (2) building national databases on hazards losses, mitigation efforts, and the social aspects of disasters; and (3) improving the use of sophisticated technology to process and evaluate risk and hazards data. As Mileti (1999) states:

> Not enough is known about the changes in or interactions among the physical, social, and constructed systems that are reshaping the nation's hazardous future. A national risk assessment should meld information from those three systems so hazards can be estimated interactively and comprehensively. . . . Local planning will require multi-hazard, community-scale risk assessment maps that incorporate information ranging from global physical processes to local resources and buildings. This information is not now available, and will require federal investment in research and risk-analysis tools and dissemination to local governments (pp. 11-12).

How did we get from one perspective to the other? Since 1975, risk assessment and hazards assessment have taken on very different meanings and conceptualizations. This has led to different, yet parallel, streams of research and the development of two distinct paradigms, hazards analysis and risk assessment, each with its own constituency, methodological approach, and vigor. These two approaches dominate environmental hazards research today but they are still not fully integrated into a comprehensive assessment of hazards and methods for reducing or mitigating the escalating costs at the local, regional, and national levels.

Hazards Paradigm

The basic underpinning of the hazards paradigm is reflected in Harlan Barrows' presidential address to the Association of American Geographers (AAG), titled "Geography as Human Ecology" (Barrows 1923). In that address, he suggested that society interacts with the physical environment and this interaction produces both beneficial and harmful effects. This relationship between people and their environment is further viewed as a series of adjustments in both the human-use and natural-events systems. Acknowledging Barrows' conceptual stance, Gilbert F. White (his student) undertook a series of floodplain studies during the 1940s and 1950s to offer a pragmatic element to Barrows' viewpoint. From these studies, the natural hazards paradigm emerged (Burton and Kates 1986). Initially, the natural hazards paradigm concentrated on five thematic areas:

1. Identification and mapping of the human occupance of the hazard zone
2. Identification of the full range of human adjustments to the hazard
3. Study of how people perceive and estimate the occurrence of hazards
4. Description of the processes whereby mitigation measures are adopted, including the social context within which that adoption takes place
5. Identification of the optimal set of adjustments to hazards and their social consequences.

A further refinement of these thematic areas led to the formulation of systems models to provide causal mechanisms for linking natural events and societal responses. Examples of these include the human adjustment to natural hazards model (Kates 1971) and the human adjustment to the risk of environmental extremes model (Mileti 1980). In addition to advancements in theory, the universality of the hazards paradigm was tested using a wide range of hazard experiences in different cultural contexts. The resulting volume (White 1994) provided detailed comparative case studies from different world regions and, today, still remains as one of the classic examples of hazard field studies.

Critics of the natural hazards paradigm (Hewitt 1983) reacted to the causal sequencing and explanations put forth about how individuals and societies adjust to hazards. Instead, they suggested that cultural, economic, political, and social forces have intensified hazards and made people more vulnerable (Blaikie et al. 1994, Kasperson et al. 1995, Hewitt 1997). More recent models place "hazards in context"—a framework that expands the traditional natural hazards perspective to include the social and political contexts within which the hazard occurs. In other words, this approach identifies factors that constrain or enable our understanding and response to hazards (Mitchell et al. 1989, Palm 1990). The contemporary hazards paradigm seeks to develop a dialogue between the physical setting, political-economic context, and the role and influence of individuals, groups, and special interests in effecting adjustments to hazards (Quarantelli 1988, Comfort et al. 1999). It does not matter where this perspective is practiced, be it sociology, geography, or the other social sciences, because the elements are essentially the same.

Risk Assessment

Most people point to Starr's (1969) seminal article on social benefit versus technological risk as the beginning of quantitative risk analysis and the development of the risk paradigm, although antecedents are found much earlier (Covello and Mumpower 1985). Quantitative risk assessment tries to define the extent of human exposure to a wide range of chemical, biological, or physical agents and the number of expected or excess deaths or increased cancers as a consequence of the use of those agents. It has also been used to assess the potential for catastrophic accidents in technological systems such as nuclear power plants (Perrow 1999).

Recognizing the need to advance research and applications in risk analysis and management, the National Science Foundation (NSF) initiated a program in 1979 to support extramural research in risk analysis and elevate the visibility of this emerging paradigm within the foundation. The early program focused on social science applications and was known as the Technology Assessment and Risk Analysis program (Golding 1992). In the early 1980s the program moved to the social and economic sciences directorate. Today, the Decision, Risk, and Management Science program at the NSF is the primary source of funding for risk analysis research.

Drawing from many fields, the interdisciplinary risk assessment field became institutionalized in 1980 with the establishment of the Society for Risk Analysis. This association includes both researchers and practitioners with interests in health, safety, and environmental risks. Their publication, *Risk Analysis*, contains articles representing the health sciences, engineering, physical sciences, and social sciences, all geared toward health and safety issues.

The formal acceptance of the risk paradigm occurred in 1983 when the National Research Council (NRC) published *Risk Assessment in the Federal Government: Managing the Process*, a report that systematized definitions and concepts as well as presenting a risk assessment framework. The risk assessment framework (still in use today) characterizes the risk paradigm. It has four primary elements: risk identification, dose-response assessment, exposure assessment, and risk characterization (NRC 1983). The NRC model was accepted as the regulatory standard under the 1980 Superfund legislation and thus became institutionalized as part of the Remedial Investigation/Feasibility process used to prioritize the cleanup of abandoned sites. Under the remedial investigations, baseline data on the site characterization (quantities and types of hazardous materials on site), the potential pathways of human exposure (inhalation, dermal, ingestion), and technical options for cleanup were required. The ultimate goal of the risk assessment process was to identify remedial options that posed the least threat to human and ecosystem health. Because of their initial development for monitoring and assessing human health risks (especially carcinogenic risks), most risk assessments use probability estimators and other statistical techniques. The results are then phrased as a 1 in 1 million chance of dying, or a 1 in 10 million chance of causing cancer in humans, or some similar metric.

A slight deviation from the traditional probabilistic risk assessment occurred in 1987, when the U.S. Environmental Protection Agency (USEPA) published *Unfinished Business* (USEPA 1987). That report compared the risks associated with a wide range of environmental problems under the USEPA's jurisdiction. Using four dimensions (human cancer risk, noncancer human health risk, ecological risk, and welfare risk), each program area was reexamined on the basis of the relative risk of each environmental problem, not just its carcinogenic potential. These risks were then rank-ordered, not defined in probabilistic terms. The relative-risk approach led to a change of emphasis within the agency—a movement away from pollution control and technological fixes and more focus on risk reduction and sustainable approaches to pollutant management. Comparative risk analysis now provides the basis for environmental policy priority setting (Davies 1996).

There was a continued broadening of the risk assessment paradigm in the early 1990s to encompass ecological risk assessments in response to pollution and ecosystem health concerns. Damages to natural systems are quantified to provide comparative data on cleanup levels and assessments of the monetary value of damages from oil spills, for example. Ecological risk assessment is the process whereby magnitudes and probabilities of an adverse effect resulting from specific human activities on particular ecosystems are determined (see Chapter 2). Throughout an ecological risk assessment, there is an assumption that discrete cause-and-effect linkages can be made either by identification of target species at risk or some other measurable surrogate that infers potential species (flora or fauna) damage.

The practice of risk assessment is fraught with methodological concerns and controversy (NRC 1994). Issues related to uncertainty in the science, variability between individuals and ecosystems, extrapolation of bioassay data to humans, and communication of risks to the public and policy makers all contribute to a lively debate among risk professionals (Dietz and Rycroft 1987). In an ideal context, risk assessment should be viewed as a process that entails an extended dialog between technical experts and interested or affected citizens (NRC 1996). In that process, it is important to get the science right and the right science, get the participation right and the right stakeholders, and develop an accurate, balanced, and informative synthesis for decision making.

Hazard and Risk Perception

One of the early issues confronting hazards researchers was to understand what people thought about natural hazards and how these perceptions influenced their choice of actions in adjusting to the sources of threats. Most of the early hazard perception work examined knowledge and attitudes about specific hazards through field studies of floods (White 1964), droughts (Saarinen 1966), and tornadoes (Sims and Baumann 1972). International perceptions were also solicited in these field studies (White 1974). These studies provided a common understanding of how people felt about extreme events: people do not recognize that they live in hazardous environments; awareness of prevention strategies makes little difference since most people know how to reduce their losses, yet the adoption of prevention measures depends on past experience with the hazard in question (Cutter 1993). While there was some input by psychologists into these early hazard perception studies (largely through research techniques), the majority were conducted by geographers.

Risk perception took a different path with more interest in the cognitive processes that give rise to the attitudes and perceptions. Early risk perception studies helped us understand biases in our judgments about risks, errors in risk estimation, and differences between experts and the general public. The psychometric paradigm developed by Paul Slovic and colleagues used an experimental approach under controlled conditions (usually college settings) to "map" risk attitudes and perceptions, suggesting that risk perceptions were quantifiable and predictable (Slovic 1987). This research provides a scientific basis for understanding why some risks are acceptable to individuals and thus society (such as smoking or air travel), whereas others are not (radioactive waste disposal, nuclear weapons fallout) (Fischhoff et al. 1978, 1979). It also helps explain why there is such a convergence between expert judgments and those of the public (Slovic 2001).

MUTUAL INTERESTS, DIVERGENT PATHS

With the overlap in interests, there is surprising little communication between hazards researchers and practitioners and the risk analysis community (White 1988). This is partly a function of the focus on extreme natural events by the hazards community, whereas the risk assessment community initially was more interested in technological risks and industrial failures. There are also methodological differences that often preclude discussion because the communities simply cannot talk to one another. Some describe this as the difference between "hard" and "soft" science. Another explanation, offered by Gilbert White (1988), suggests that risk analysis fails to include the social structure or social context within which those risks occur, a critical element for hazards researchers. There are a number of prominent social science researchers who have tried to bridge the gap between the two perspectives, but with limited success.

Despite the professional segregation, many similarities exist between the risk assessment and hazard analysis paradigms (Table 1-1). For example, hazard analysis has three components: identification, assessment, and management/mitigation. Within this, hazard identification is concurrent with the mapping of hazard zones; assessment is the determination of the vulnerability and the potential population at risk, including their socioeconomic characteristics; and hazards management/mitigation

TABLE 1-1 Risk Assessment and Hazard Analysis Paradigms

Elements	Risk Assessment	Hazard Analysis
Hazard identification	Does the agent/toxin cause the adverse effect? Chemical Y has a 1 in 1 million chance of causing cancer in humans.	What is the threat? What is the occurrence of the hazard? Mapping of specific hazards and/or hazard zones
Dose-response assessment	What is the relationship between dose and incidence in humans? Exposure to *X* parts per million of chemical *Y* for a period of 2 days causes liver damage.	What are the magnitude, frequency, and duration of the event? What are the potential human consequences of the event?
Exposure assessment	What exposures are currently experienced or anticipated under different conditions? How much of the toxin will reach a targeted population or how many people will receive some exposure?	What is the pattern of human occupance in hazard zones? What is the vulnerability of people and places to hazards?
Risk characterization	What is the estimated incidence of the adverse effect in a given population? What is the likelihood that an agent of concern will be realized in exposed people?	What accounts for different adjustments and adaptations to hazards? How do societies prepare for, mitigate, and respond to risks and hazards?

includes the range of options or adjustments society is willing to take to respond to hazards and disasters. Conceptually then, many of the same questions are addressed in both perspectives, but the methodologies used to respond to the queries are radically different.

One of the major obstacles to the integration of the risk and hazards paradigms is this methodological divide and the exclusive use of a reductionist analytical framework found in risk analysis. The heavy reliance on quantitative methods and models often excludes people as dynamic factors. On the other hand, the hazards paradigm must move beyond a simple descriptive methodology (quantitative or qualitative) to a more integrated analytical framework that permits the assessment of larger and more complex databases and more robust empirical field studies.

MOVING FROM THEORY TO PRACTICE

Translation of the results from researchers in hazards and risk analysis to the actual practice of risk reduction and hazard mitigation is essential. Risk analysis is most often used in regulatory standard setting or rule making (Hamilton and Viscusi 1999). Hazards assessment is more likely to be used in planning or programmatic contexts. In this regard, both have very different constituencies, but equally share a commitment for reducing societal risks and hazards. Several national and international efforts have tried to increase awareness of hazards and reduce their impacts.

In 1994, the U.S. Federal Emergency Management Agency (FEMA) released its National Mitigation Strategy in the hopes of reversing the escalating losses from natural hazards through a combination of private-public partnerships and incentives for local communities. By creating "disaster-resistant communities" under its Project Impact program, FEMA hoped to reduce the impact of hazard events on people and places through improved mitigation. It is too soon to gauge the effectiveness of this program, but the number of communities who have bought into the idea is increasing exponentially.

The International Decade for Natural Disaster Reduction (IDNDR) has just concluded. This United Nations program focused worldwide attention on the increasing losses from natural hazards and urged member nations to implement actions to reduce losses in their country. One of the outcomes of the decade was a clear shift toward mitigation rather than simple response and recovery (Press and Hamilton 1999). According to the NRC's Board on Natural Disasters (BOND 1999), future U.S. efforts should focus on the following high-priority areas:

- Improved risk assessment
- Implementation of mitigation strategies such as land-use planning, building codes, tax incentives, and infrastructure improvements
- Improved warning technologies and their dissemination and use
- Improved use of insurance for rewarding risk reduction behavior
- Assistance to disaster-prone developing nations.

Note that the recommendations from the IDNDR are largely structural in nature and do not call for any major philosophical shifts in understanding the patterns of development, their impact on society, or how these practices contribute to hazard vulnerability. It offers a series of options that ultimately will postpone or redistribute risks and hazards, not necessarily reduce them—more akin to "business as usual."

In contrast, the Second Assessment calls for an understanding of the underlying social processes that give rise to hazards in the first place. In moving toward a more sustainable future, "a locality can tolerate—and overcome—damage, diminished productivity, and reduced quality of life from an extreme event without significant outside assistance" (Mileti 1999:5-6). To achieve sustainability through mitigation, the following are required:

- Maintain and enhance environmental quality
- Maintain and enhance people's quality of life
- Foster local resiliency and responsibility
- Support a strong local economy by using mitigation actions that do not detract from the economy
- Ensure inter- and intra-generational equity in the selection of mitigation options
- Adopt local consensus building.

Although the Second Assessment sets out admirable goals, many of these may not be realistic or achievable, given contemporary political and economic priorities. On the other hand, there may be no better place to begin the process of restructuring nature-society interactions.

CONCLUSION

As we have seen, not only are there divergent paths within the hazards and risk research communities, but there are also differing approaches to reducing losses domestically and internationally. With this chapter as a backdrop, we can now turn our attention to examining the evolution of American hazardscapes as we try to understand the variability in and delineation of hazard-resistant or disaster-prone places. We begin with an overview of vulnerability and hazards assessment—what these concepts mean and how we measure them.

CHAPTER 2

Methods for Determining Disaster Proneness

Arleen A. Hill and Susan L. Cutter

Even with improvements in detection and warning systems, the direct losses associated with hazard events has steadily risen during the past three decades (van der Wink et al. 1998). Why is it that some places appear to be more disaster-prone whereas other communities seem to be somewhat immune from the impact of natural hazards? What makes some places more vulnerable to natural hazards than others? Is it that some communities are simply more at risk, or they have more people who lack adequate response mechanisms when the disaster strikes, or is it some combination of the two? This chapter reviews some of the contemporary hazard assessment tools and techniques that help us to understand societal vulnerability to hazards.

VULNERABILITY AND THE POTENTIAL FOR LOSS

In its simplest form, vulnerability is the potential for loss. Like the term sustainability, vulnerability means different things to different people. For example, in the summary volume of this series, Mileti (1999) offered that vulnerability is "the measure of the capacity to weather, resist, or recover from the impacts of a hazard in the long term as well as the short term" (p. 106). Vulnerability has been variously defined as the threat of exposure, the capacity to suffer harm, and the degree to which different social groups are at risk (Cutter 1996a). All are consistent with the more general definitions provided here. Perhaps equally important is the notion that vulnerability varies by location (or space) and over time—it has both temporal and spatial dimensions. This means that vulnerability can be examined from the community level to the global level, can be compared from place to place, and can be studied from the past to the present and from the present to the future. Most important to remember is that geography matters when discussing the vulnerability of people and places to environmental hazards.

Types of Vulnerability

There are many types of vulnerability of interest to the hazards community, but three are the most important: individual, social, and biophysical. Individual vulnerability is the susceptibility of a person or structure to potential harm from hazards. Scientists and practitioners from engineering, natural sciences, and the health sciences are primarily interested in this type of vulnerability. The structural integrity of a building or dwelling unit and its likelihood of potential damage or failure from seismic activity are examples of vulnerability at the individual-level scale. Building codes, for example, are designed to reduce individual structural vulnerability. Another example comes from the health sciences and focuses on the vulnerability of individual people, the potential exposure of the elderly to heat stress during the summer. In both instances, the characteristics of the individual structure (building materials, design) or person (age, diet, smoking habits, living arrangements) largely dictate their degree of vulnerability. The primary unit of analysis is an individual person, structure, or object.

On a more general level, we have social vulnerability, which describes the demographic characteristics of social groups that make them more or less susceptible to the adverse impacts of hazards. Social vulnerability suggests that people have created their own vulnerability, largely through their own decisions and actions. The increased potential for loss and a reduction in the ability to recover are most often functions of a range of social, economic, historic, and political processes that impinge on a social group's ability to cope with contemporary hazard events and disasters. Many social scientists working in the field today, especially those working with slow-onset hazards (drought, famine, hunger) or those working in developing world contexts, use the social vulnerability perspective. Some key social and demographic characteristics influencing social vulnerability include socioeconomic status, age, experience, gender, race/ethnicity, wealth, recent immigrants, tourists and transients (Heinz Center 2000a).

Biophysical vulnerability, the last major type, examines the distribution of hazardous conditions arising from a variety of initiating events such as natural hazards (hurricanes, tornadoes), chemical contaminants, or industrial accidents. In many respects, biophysical vulnerability is synonymous with physical exposure. The environmental science community mostly addresses issues of biophysical vulnerability based on the following characteristics of the hazards or initiating events: magnitude, duration, frequency, impact, rapidity of onset, and proximity. These types of studies normally would produce statistical accountings (or in some instances maps) that delineate the probability of exposure—that is, areas that are more vulnerable than others such as 100-year floodplains, seismic zones, or potential contamination zones based on toxic releases.

The integration of the biophysical and social vulnerability perspectives produces the "hazards of place" model of vulnerability (Cutter 1996a, Cutter et al. 2000). In our view, understanding the social vulnerability of places is just as essential as knowing about the biophysical exposure. The integrating mechanism is, of course, place. These places (with clearly defined geographic boundaries) can range from census divisions (blocks, census tracts), to neighborhoods, communities, counties, states, regions, or nations. Among the many advantages of this approach is that it permits us to map vulnerability and compare the relative levels of vulnerability from one place to another or from region to region and, of course, over time. In this way, we have a very good method for differentiating disaster-prone from disaster-resilient communities, identifying what suite of factors seem to influence the relative vulnerability of one place over another, and monitoring how the vulnerability of places changes over time as we undertake mitigation activities.

Developing Risk, Hazard, and Vulnerability Assessments

As mentioned in Chapter 1, the terms *risk* and *hazard* have slightly different meanings. It should come as no surprise that, in developing risk or hazards assessments, there are subtle differences in meaning and approaches as well. Risk assessment, is a systematic characterization of the probability of an adverse event and the nature and severity of that event

(Presidential/Congressional Commission on Risk Assessment and Risk Management 1997). Risk assessments are most often used to determine the human health or ecological impacts of specific chemical substances, microorganisms, radiation, or natural events. Risk assessments (the relationship between an exposure and a health outcome) normally focus on one type of risk (e.g., cancers, birth defects) posed by one substance (e.g., benzene, dioxin) in a single media (air, water, or land). In the natural-hazards field, risk assessment has a broader meaning, and involves a systematic process of defining the probability of an adverse event (e.g., flood) and where that event is most likely to occur. Much of the scientific work on modeling, estimating, and forecasting floods, earthquakes, hurricanes, tsunamis, and so on, are examples of risk assessments applied to natural hazards (Petak and Atkisson 1982).

Vulnerability assessments include risk/hazard information, but also detail the potential population at risk, the number of structures that might be impacted, or the lifelines, such as bridges or power lines (Platt 1995), that might be damaged. Vulnerability assessments describe the potential exposure of people and the built environment. The concept of vulnerability incorporates the notion of differential susceptibility and differential impacts. Exposures may not be uniformly distributed in space, nor is the societal capacity to recover quickly the same in all segments of the population. It is the exposure to hazards and the capacity to recover from them that define vulnerability. Thus, vulnerability assessments must incorporate both risk factors and social factors in trying to understand what makes certain places or communities more susceptible to harm from hazards than others. This makes vulnerability assessments more difficult to undertake than simple risk analyses because they require more data (some of which may not be available) and have more complex interactions that need careful consideration.

For our purposes, however, we use the terms risk assessment and hazards assessment interchangeably. There is a rich literature on risk/hazards assessments and vulnerability assessments that range from very localized experiences to the development of global models of vulnerability (Ingleton 1999). We describe a number of these approaches in the following section.

METHODS OF ASSESSMENT

There have been many notable advances in hazards and risk modeling during the past two decades. These are described in more detail in the following subsections.

Risk Estimation Approaches and Models

The majority of risk estimation models are hazard specific. Some are designed for very general applications, such as the hurricane strike predictions issued by the National Hurricane Center (NHC), whereas others are designed to aid in the selection of specific protective actions based on the plume from an airborne chemical release. It is impossible to adequately review all of the existing risk estimation models. Instead, we highlight a number of the most important and widely used within the hazards field. A more detailed description of these models can be found in Appendix A.

Hazardous Airborne Pollutant Exposure Models

Dispersion models estimate the downwind concentrations of air pollutants or contaminants through a series of mathematical equations that characterize the atmosphere. Dispersion is calculated as a function of source characteristics (e.g., stack height, rate of pollution emissions, gas temperature), receptor characteristics (e.g., location, height above the ground), and local meteorology (e.g., wind direction and speed, ambient temperature). Dispersion models only capture the role of the atmosphere in the delivery of pollutants or an estimation of the potential risk of exposure. They do not provide an estimate of the impacts of that exposure on the community or individuals because those specific environmental and individual factors are not included in the model. They are crude representations of airborne transport of hazardous materials and lead to over- or underestimation of concentrations (Committee on Risk Assessment of Hazardous Air Pollutants 1994).

Many of the dispersion models currently in use, such as the U.S. Environmental Protection Agency's (USEPA) Industrial Source Complex (ISC), can accommodate multiple sources and multiple receptors. Another aspect of these models is the ability to include short-term and long-term versions, which can model different timing and duration of releases (USEPA 1995). Regulators, industries, and consultants commonly use ISCST3, the short-term version. In addition to permitting and regulatory evaluations, dispersion models have been applied to studies of cancer risk from urban pollution sources (Summerhays 1991). At a National Priority List landfill in Tacoma, Washington, the ISCST3 model was used to predict the maximum ground-level concentrations of volatile organic compounds vented from the contaminated site (Griffin and Rutherford 1994).

The most important air dispersion model for emergency management is the Areal Locations of Hazardous Atmospheres (ALOHA)/Computer-Aided Management of Emergency Operations (CAMEO) model of airborne toxic releases, developed by the USEPA. The ALOHA model uses a standardized chemical property library as well as input from the user to model how an airborne release will disperse in the atmosphere after an accidental chemical release (USEPA 2000a). Currently, the model is used as a tool for response, planning, and training by government and industry alike. Graphical output in the form of a footprint with concentrations above a user-defined threshold can be mapped (Figure 2-1) using a companion application (MARPLOT). CAMEO is a system of software applications used to plan for and respond to chemical

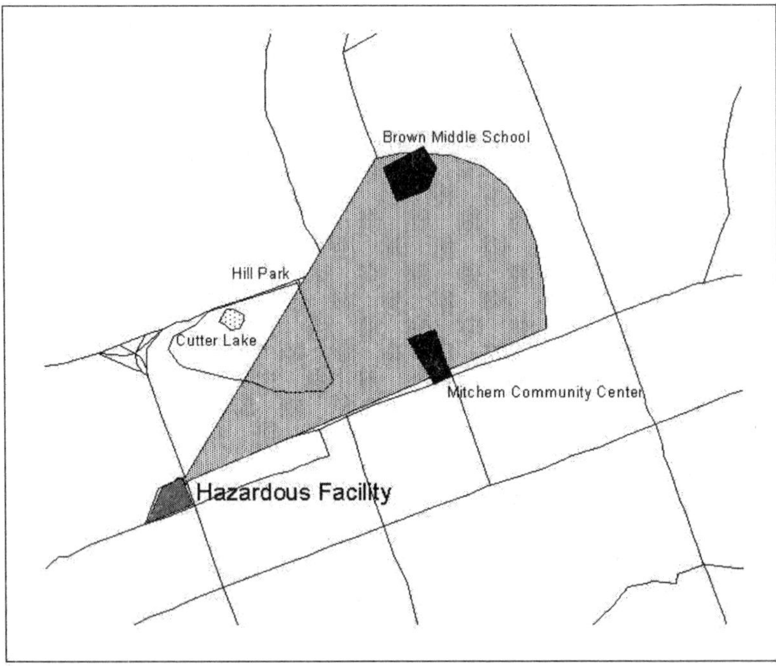

FIGURE 2-1 Stylized version of the computer output from the ALOHA model, showing the plume path from an airborne toxic release. See USEPA 2000b, *http://response.restoration.noaa.gov/cameo/aloha.html* for other examples.

emergencies. Developed by the USEPA's Chemical Emergency Preparedness and Prevention Office (CEPPO) and the National Oceanic and Atmospheric Administration (NOAA), it is designed to aid first-responders with accurate and timely information. Integrated modules store, manage, model, and display information critical to responders. CAMEO has a database of response recommendations for 4,000 chemicals and works with the ALOHA air dispersion model and MARPLOT mapping module to provide firefighting, physical property, health hazard, and response recommendations based on the specific chemical identified (USEPA 2000b).

Storm Surge

Potential storm surge inundation of coastal areas is determined by the Sea, Lake, and Overland Surges (SLOSH) model developed by the U.S. National Weather Service (NWS) in 1984 (Jelesnianski et al. 1992). As a two-dimensional, dynamic, numerical model, SLOSH was developed initially to forecast real-time hurricane storm surges. SLOSH effectively computes storm surge heights, where model output is a maximum value for each grid cell for a given storm category, forward velocity, and landfall direction. SLOSH is used primarily in pre-impact planning to delineate potential storm surge inundation zones and can be repeated for different hurricane scenarios at the same location (Garcia et al. 1990).

NOAA recommends that emergency managers use two slightly different model outputs in their evacuation planning—MEOW and MOM. MEOW is the Maximum Envelope of Water and represents a composite of maximum high-water values per grid cell in the model run. The National Hurricane Center (NHC) combines the values for each grid cell and then generates a composite value for a specific storm category, forward velocity, and landfall direction. Of more use to emergency managers is the MOM (Maximum of the MEOW), a composite measure that is the maximum of the maximum values for a particular storm category (NOAA 2000a). In other words, this is the worst-case scenario. SLOSH model output is in the form of digital maps that show the calculated storm surge levels as a series of contours or shaded areas (Figure 2-2, see color plate following page 22). These maps form the basis for local risk estimates upon which evacuation plans are developed. Comparisons of the SLOSH real-time forecasts and actual observations of storm surge confirm that this model is extremely useful to emergency management officials at local, state, and national levels. This output provides not only an important pre-impact planning tool (locating shelters and evacuation routes), but also contributes to recovery and mitigation efforts in coastal communities (Houston et al. 1999). The model also has been utilized in the revision of coastal flood insurance rate maps (FIRMS).

Regional SLOSH model coverage includes the entire Gulf and Atlantic coastlines of the United States and parts of Hawaii, Guam, Puerto Rico, and the Virgin Islands. Modeling of SLOSH basins has been extended internationally to include the coastal reaches of the People's Republic of China and India (NOAA 2000a).

Hurricane Strike Forecasting and Wind Fields

Both long-range and storm-specific forecasting and modeling efforts exist for hurricanes. The NHC provides forecast information on storm track, storm intensity, and surface winds for individual storms. The long- or extended-range forecasts categorize or predict the activity of a specific basin over a specific season. Predictions or estimations for the Atlantic Basin are based on statistical models and the experience of a forecasting team. The statistical model incorporates global and regional predictors known to be related to the Atlantic Basin hurricane season, mostly derived from historical data. The model is run and then qualitatively adjusted by the forecast team based on supplemental information not yet built into the model (Gray et al. 1999).

Advances in weather satellites, forecasting models, research, and empirical data have led to a reduction of errors in forecasted path by 14 percent in the past 30 years (Kerr 1990). Although some of the statistical models are being phased out, the more dynamic models taking their place, such as GFDL, UKMET, and NOGAPS (see Appendix A), allow the NHC to develop an average of these storm-track predictions and then send appropriate warning messages. Although this approach has led to a decrease in errors in strike probabilities (180 miles error at 48 hours out; 100 miles error at 24 hours out) it has resulted in an increase in the amount of coastline that must be warned per storm (Pielke 1999). There are a number of explanations for this: (1) the desire to base evacuation decisions on the precautionary principle and develop a very conservative estimate of landfall based on official NHC forecasts; (2) the westward or left bias of many of the models, thus the need to warn a larger area to avoid any last-minute change in the track; and (3) larger coastal populations requiring longer evacuation times.

At the same time, improvements in cyclone intensity modeling (SHIFOR, SHIPS, CLIPER; see Appendix A) allow forecasters to improve their estimation of the intensity of these systems by 5-20 knots. This type of modeling provides better estimates of wind fields at the surface and near surface. With improved air reconnaissance and the use of Global Positioning System drop-windsondes (Beardsley 2000), we now have a better understanding of wind patterns not only at the surface but throughout the entire storm system. Additional information on hurricane forecasting models can be found in Appendix A and at NOAA's Web site (NOAA 2000a).

Tornado Risk Estimation

The issuance and communication of tornado watches and warnings is a vital part of protecting lives and property from severe storms. A complex network that relies on both sophisticated equipment and trained observers monitors the development of severe storm cells. NOAA's National Severe Storms Forecast Center (NSSFC) in Kansas City, Missouri, the NWS SKYWARN System, radar and trained spotters, and the NOAA National Severe Storms Laboratory (NSSL) in Norman, Oklahoma are part of the complex network.

In partnership with the NWS, the NSSL generates forecasts based on numerical weather prediction models (see Appendix A). These models provide temperature, pressure, moisture, rainfall, and wind estimates as well as geographically tracking severe storms and individual storm cells.

Doppler radar provides estimates of air velocity in and near a storm and allows identification of areas of rotation. Detection of potentially strong thunderstorm cells (also called mesocyclone signatures) can be identified in parent storms as much as 30 minutes before tornado formation. Signatures of the actual tornadic vortex can also be observed on Doppler radar by the now-distinct hook-shaped radar echo (Figure 2-3, see color plate following page 22). The signature detection process remains subjective, however, and thus prone to human errors in image interpretation, especially in real time. Some tornado-generating storms have typical signatures, but others do not. Also, some storms with these mesocyclone signatures never generate tornadoes. Nevertheless, Doppler radar has helped increase the average lead time for risk estimation and the implementation of tornado warnings from less than 5 minutes in the late 1980s to close to 10 minutes today (Anonymous 1997).

Flood Risk

Estimating flood risk depends on the type of flooding: coastal (storm surge and tsunamis), riverine (river overflow, ice-jam, dam-break), and flash floods (with many different subcategories) (IFMRC 1994). Coastal flooding risk due to storm surge (see storm surge discussion p. 19) is developed by a slightly different set of models than riverine or flash floods. The complex relationship between hydrological parameters such as stream channel width, river discharge, channel depth, topography, and hydrometeorological indicators of intensity and duration of rainfall and runoff produces estimates of the spatial and temporal risk of flooding.

Working in tandem, the U.S. Geological Survey (USGS) and the NWS provide relevant risk data for flood events. The NWS has the primary responsibility for the issuance of river forecasts and flood warnings through its 13 Regional River Forecast Centers. The USGS provides data on river depth and flow through its stream gauge program, resulting in real-time river forecasting capability. In fact, real-time river flow data for gauged streams are now available on the World Wide Web for anyone to use (USGS 2000a). Flow or discharge data are more difficult to measure accurately and continuously, and so, hydrologists employ rating curves, which are pre-established river stage/discharge relations that are periodically verified by field personnel.

NOAA's Hydrometeorological Prediction Center (HPC) is another important partner in determining flood risk. The HPC provides medium-range (3-7 days) precipitation forecasts as well as excessive rainfall and snowfall estimates. Through the use of quantitative precipitation forecasts, forecasters can estimate expected rainfall in a given basin and the accumulated precipitation for 6-hour intervals. This information is transmitted to the NWS River Forecast Centers and is also available in real time (HPC 2000). Using river stage, discharge, and rainfall, hydrologic models are employed to see how rivers and streams respond to rainfall and snowmelt. These modeled outputs provide the risk information that is transmitted to the public—height of the flood crest, date and time that river is expected to overflow its banks, and date and time that the river flow is expected to recede within its banks (Mason and Weiger 1995).

Coastal Risk

The Coastal Vulnerability Index evaluates shoreline segments on their risk potential from coastal erosion or inundation (Gornitz et al.

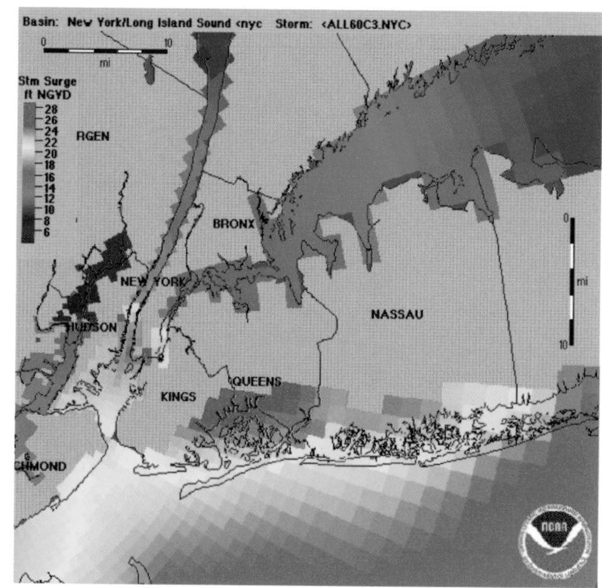

FIGURE 2-2 SLOSH model output for a Category 3 hurricane along Long Island, New York. Source: *http://hurricane.noaa.gov/prepare/longislandc3.htm.*

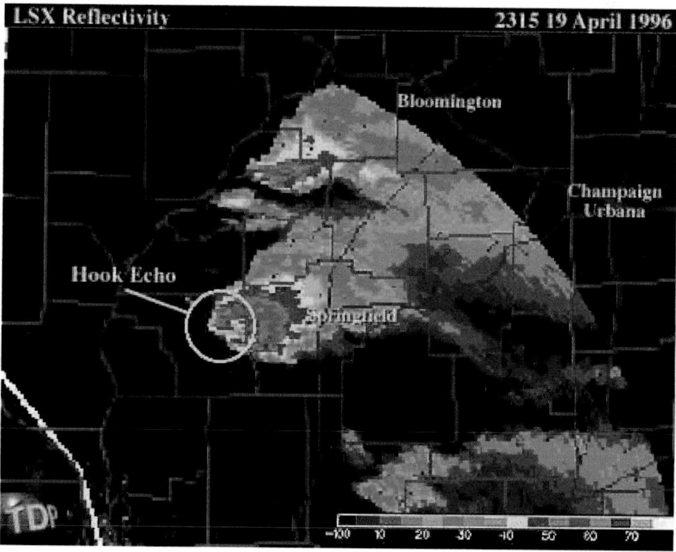

FIGURE 2-3 Doppler radar image of the April 1996 Jacksonville, Illinois, tornado, illustrating the diagnostic hook echo. Source: *http://ww2010.atmos.uiuc.edu/(Gh)/arch/cases/960419/nxrd/jack.rxml.*

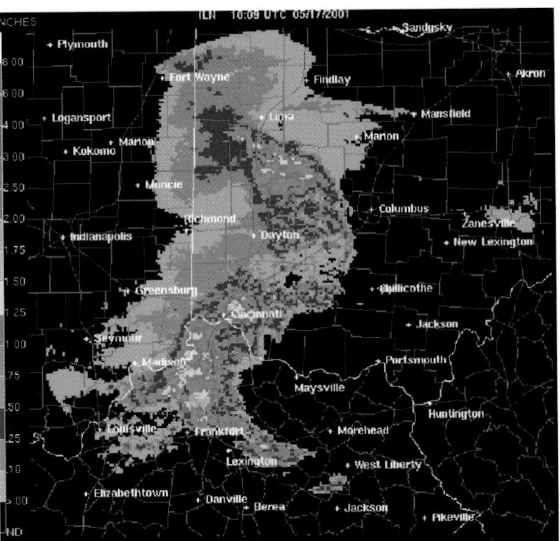

FIGURE 3-3 Rainfall estimates from Doppler radar for severe thunderstorms in southwestern Ohio and northern Kentucky, May 17, 2001. Source: *http://weather.noaa.gov/radar/latest/DS.78ohp/si.kiln.shtml.*

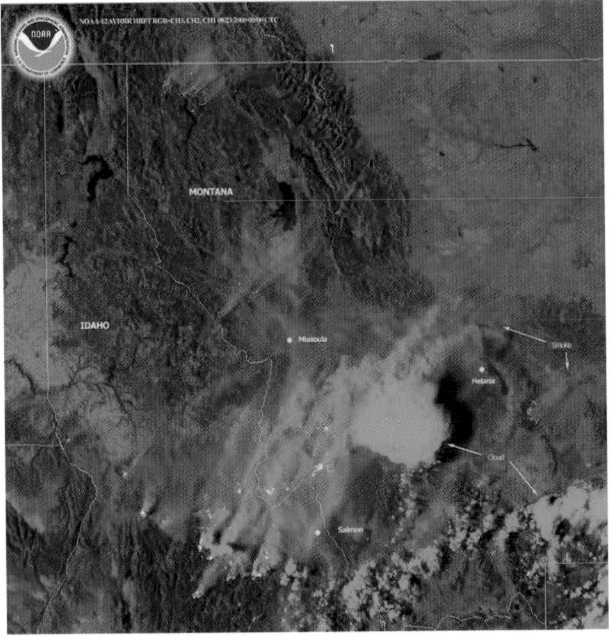

FIGURE 3-4 A remotely sensed image of heat signatures (red) and smoke (blue) from fires burning in the western United States on August 23, 2000. Source: *http://www.osei.noaa.gov/Events/Current.*

1994, FEMA 1997a). High-risk coastlines are defined by low coastal elevations, histories of shoreline retreat, high wave/tidal energies, erodible substrates, subsidence experience, and high probabilities of hurricane and/or tropical storm hits. The index uses 13 biophysical variables such as elevation, wave heights, hurricane probability, and hurricane intensity, which are ranked from low to high (1 to 5). Three different indicators were developed (permanent inundation, episodic inundation, and erosion potential) from these 13 variables and then weighted to produce an overall score for each of the 4,557 U.S. shoreline segments examined. The data are geocoded and can be used in conjunction with other geographic information to produce vulnerability assessments.

A recent report completed by the Heinz Center (2000b) delineates the potential erosion hazard for selected study segments of U.S. coastlines from Maine to Texas, southern California to Washington, and along the Great Lakes. Using historic shoreline records dating back to the 1930s or earlier, historic rates of erosion are used to calculate an annual erosion rate. These rates of erosion are then mapped using the current shoreline as the base, with projections to 30 years and 60 years. The future rates of erosion are conservative estimates. For example, there is no consideration of accelerated sea-level rise due to increased global warming or increased development along the coastline, both of which might accelerate erosion rates.

Another approach to coastal risk examines the potential hazards of oil spills on coastal environments. The Environmental Sensitivity Index (ESI) is used to identify shoreline sensitivity to oil spills based on the nature of the biological communities, sediment characteristics, and shoreline characteristics (including both physical attributes and cultural resources) (Jensen et al. 1993, 1998). Using remote sensing to monitor changes in coastal wetland habitats and a geographic information system to catalog, classify, and map sensitive areas, the ESI has been applied not only to coastlines, but to tidal inlets, river reaches, and regional watersheds.

Seismic Risk

Seismic risks are delineated in two different ways—through the identification of specific known surface and subsurface faults and through the regional estimation of ground motion expressed as a percentage of the peak acceleration due to gravity. Both rely on historical seismicity, using the past as a key to future earthquake activity. Assessments that yield maps of active faults or data on historical seismic activity are the simplest forms of seismic risk assessment. However, probabilistic maps that incorporate the likelihood of an event or exceedence of some ground-shaking threshold are now being generated and increasingly used as estimators of seismic risk.

Two approaches used in assessing an earthquake hazard are probabilistic and deterministic methods. The probabilistic approach attempts to describe the integrated effects from all possible faults at an individual site. These assessments recognize uncertainties in our knowledge of fault parameters, earthquake magnitude and intensity, and the responses of built structures. The probabilistic method requires the user to define the level of risk for consideration, thus introducing the concept of acceptable risk and the consideration of critical facilities (Yeats et al. 1997). Probabilistic studies can be further divided into time-dependent and time-independent models. Recent examples include the Working Group 2000 report on the San Francisco Bay Area (time-dependent) (USGS 2000b) and the national seismic hazard maps (time-independent) (Frankel et al. 1997, USGS 2000c).

Deterministic methods specify a magnitude or level of ground shaking to be considered. Often, a single fault is considered and seismic parameters for a "maximum credible" event are applied. The probability of that earthquake occurring is not incorporated in the analysis. Instead, three types of information are used: historical and instrumental record, evidence and physical parameters from seismogenic faults, and/or paleoseismic evidence of prehistoric earthquakes (Yeats et al. 1997). These types of models commonly represent a "worst-case scenario," or the maximum risk people or a particular place could be exposed to.

USGS maps provide estimates of the probability of exceeding certain levels of ground motion in a specified time (e.g., 10 percent probability of exceedence in 50 years). Time periods range from 100 to 2,500 years (Algermissen and Perkins 1976, Algermissen et al. 1982). National seismic hazard maps have recently been generated (Frankel et al. 1996) and have been used to develop county-level seismic risk assessments based on the default soil site conditions (Nishenko 1999, FEMA 2000a). These seismic maps are now used in the National Earthquake Hazards Reduction Program's (NEHRP) guidance for hazards loss reduction in new and existing buildings.

Vulnerability Assessments

The science of vulnerability assessments is not nearly as advanced as for risk estimation. In fact, vulnerability science is really in its infancy. Whether it is an analysis of the potential physical and economic impacts of climate change on climate-sensitive sectors such as agriculture, water resources, and the like (U.S. Country Studies Program 1999) or the integration of vulnerability into sustainable development programs (OAS 1991), the issue remains the same. What indicators do we use to measure vulnerability and how do we represent that information to decision makers?

More often than not, vulnerability indicators are single variables (lifelines, infrastructure to support basic needs, special-needs populations, schools), but occasionally multidimensional factors such as food aid, social relations, and political power have been incorporated (Blaikie et al. 1994). The research on vulnerability assessments is characterized by case studies ranging from very localized analyses of a city (Colten 1986, 1991), to a county or state (Liverman 1990), to more regional perspectives (Downing 1991, Lowry et al. 1995) and by a focus on a single hazard (Shinozuka et al. 1998). There are a variety of methods used to determine vulnerability, and many include some form of geographic information system (GIS) and maps that convey the results.

Activities within the Organization of American States (OAS) provide some of the most innovative examples of the role that hazard vulnerability plays in international development assistance. Through its Working Group on Vulnerability Assessment and Indexing, OAS is trying to develop a common set of metrics for measuring vulnerability to natural hazards that could then be used in disaster preparedness and response efforts and in financing disaster reduction among member nations. The most advanced efforts in assessing vulnerability are through the Caribbean Disaster Mitigation Project (OAS 2000).

Within the United States, hazard identification and risk assessment are among the five elements in FEMA's (1995) National Mitigation Strategy. Project Impact has called for the undertaking of hazard identification (risk assessment) and vulnerability assessment, yet it provides very little technical guidance on how to conduct such analyses other than consulting with FEMA or developing partnerships with professional associations. Despite this lack of direction, a number of states have adopted innovative tools for vulnerability assessments. For example, Florida utilized a commercial product, the Total Arbiter of Storms (TAOS) model, developed by Watson Technical Consulting (2000) (and distributed through Globalytics). This model simulates the effect of selected hazards (waves, wind, storm surge, coastal erosion, and flooding) as well as their impact on both physical and built environments including damage and economic loss estimates. Aspects of the model can also be used for real-time hurricane tracking, track forecasting, and probabilistic modeling of hurricane and tropical cyclone hazards. The TAOS hazard model has also been used by a number of Caribbean nations to assess the risk of storm surge, high wind, and wave hazards and their local impact (OAS 2000). Florida has also developed a manual for local communities, which uses a GIS-based approach to risk mapping and hazards assessment, including specific guidance on hazard identification and vulnerability assessment (State of Florida 2000).

Single-Hazard Vulnerabilities

There are a number of noteworthy efforts to determine vulnerability using single hazards. The most advanced are the U.S. Agency for International Development's (U.S. AID) Famine Early Warning System (FEWS) and the United Nations Food and Agricultural Organization's Africa Real Time Environmental Monitoring Information System (ARTEMIS) (Hutchinson 1998). Both of these monitoring programs, which include remote sensing and GIS, are part of the Global Information and Early Warning System (GIEWS) and provide data for vulnerability assessments that evaluate national and international food security issues. The vulnerability assessment is used to describe the nature of the problem and classifies, both qualitatively and quantitatively, who is affected, the impacted area, and potential interventions. These assessments are done at a variety of levels from the household to the national level. Finally, a food security and vulnerability profile can be generated that gives a historic analysis of food availability and access. This profile also provides a more comprehensive view of the level, trends, and factors that influence food security (or insecurity) for individual population groups or nations. These profiles can also be mapped to illustrate their vulnerability geographically (Figure 2-4).

Another example of a hazard-specific vulnerability assessment was conducted by NOAA's Coastal Services Center for hurricane-induced coastal hazards in Alabama (NOAA Coastal Services Center NDa). In addition to that product, they have also produced a GIS-based assessment tool that defines risk areas, identifies critical infrastructure, and

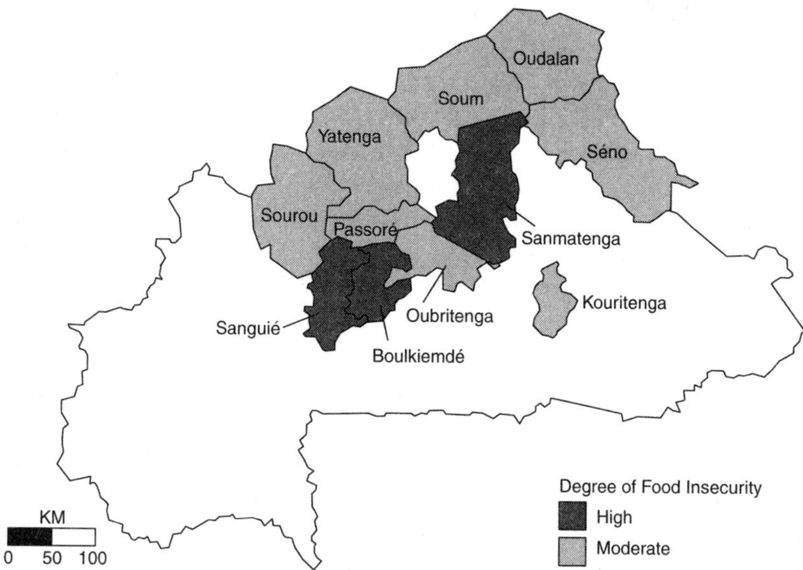

FIGURE 2-4 Food insecurity in Burkina Faso (1999-2000) based on U.S. AID's FEWS current vulnerability assessment for March 2000. Source: *http://www.fews.org/va/vapub.html.*

maps the potential impact from coastal hazards. Using a case study of coastal hazards in New Hanover County, North Carolina, the tool provides a tutorial on how to conduct vulnerability assessments and develop priorities for hazard mitigation opportunities (NOAA Coastal Services Center NDb).

Multihazard Approaches

A number of multihazard vulnerability assessments have been conducted, but all are quite localized in scale. For example, Preuss and Hebenstreit (1991) developed a vulnerability assessment for Grays Harbor, Washington. Risk factors included primary and secondary impacts from an earthquake, including a tsunami event and toxic materials release. Social indicators included land use and population density and distribution. Another example is the county-level work by Toppozada et al. (1995) on Humboldt and Del Norte, California. Mapping the risk information (tsunami waves, ground failure, fault rupture, liquefaction, and landslides) and societal impacts (buildings, infrastructure, lifelines) allowed for the delineation of vulnerable areas within the counties. The work by the USGS on Hurricane Mitch and its devastating impact in Central America is another example (USGS 1999).

Lastly, the work of Cutter and colleagues (Mitchell et al. 1997, Cutter et al. 2000) in Georgetown County, South Carolina, provides one of the most comprehensive methodologies for county-level vulnerability assessments. Utilizing both biophysical risk indicators (e.g., hurricanes, seismicity, flooding, hazardous materials spills) and social vulnerability indicators (population density, mobile homes, population over 65, race) within a GIS allows for determining the geographic distribution of vulnerability within the county. Biophysical risk was determined by the historic frequency of occurrence of hazard events in the county and applied to the specific impacted areas (e.g. 100-year floodplain for flood occurrence; SLOSH-model inundation zones for hurricanes). Areas of high social vulnerability can be examined independently of those areas of high biophysical risk. They can also be examined together to get an overall perspective of the total vulnerability of the country to environmental hazards. More importantly, this methodology enables the user to examine what specific factors are most influential in producing the overall vulnerability within the county.

Exposure Assessment and Loss Estimation Methods

Loss estimates are important in both pre-impact planning and in post-disaster response, yet we have very little systematic data on what natural hazards cost this nation on a yearly basis. In addition, there is no standardized estimation technique for compiling loss data from individual events, or any archiving system so that we can track historic trends (NRC 1999a, b). Although we are improving our data collection on biophysical processes (risk and vulnerability), some have argued that our data on natural hazard losses resembles a piece of Swiss cheese—a database with lots of holes in it.

There are many different types of loss estimation techniques that are available. Their use largely depends on what types of losses one is interested in measuring (economic, environmental) and the scale (individual structures or entire county or nation) of estimation.

HAZUS

FEMA, in partnership with the National Institute for Building Sciences (NIBS), developed HAZUS (Hazards US). This is a tool that can be used by state and local officials to forecast damage estimates and economic impacts of natural hazards in the United States. At present, HAZUS includes the capability to use both deterministic and probabilistic earthquake information for loss estimation. Although the earthquake loss estimation module is the only one available at the moment, HAZUS eventually will include a wind-loss component and a flood-loss module (scheduled for release in 2002/2003).

In HAZUS, local geology, building stock and structural performance, probabilistic scenario earthquakes, and economic data are used to derive estimates of potential losses from a seismic event. A modified GIS displays and maps the resulting estimates of ground acceleration, building damage, and demographic information at a scale determined by the user (e.g., census tract, county). Designed for emergency managers, planners, city officials, and utility managers, the tool provides a standardized loss estimate for a variety of geographic units. Generally speaking, four classes of information are provided: (1) map-based analyses (e.g., potential ground-shaking intensity), (2) quantitative estimates of losses (e.g., direct recovery costs, casualties, people rendered homeless), (3) functional losses (e.g. restoration times for critical facilities), and (4) extent of induced hazards (e.g., distribution of fires, floods, location of hazardous materials). HAZUS calculates a probable maximum loss. It also calculates average annual loss, a long-term average that includes the effects of frequent small events and infrequent larger events (FEMA 2000a).

At present, losses are generated on the basis of "scenario earthquakes," which is a limitation of the tool because the location and magnitude of the "scenario earthquakes" may not represent the actual magnitude or location of future events. Default data describe geology, building inventory, and economic structure in general terms and can be used to produce very generalized loss estimates on a regional scale. However, communities must supplement these general data with local-level data in order to assess losses for individual communities such as cities, towns, and villages. Unfortunately, these local-level data often are unavailable. Despite these limitations, HAZUS does enable local communities to undertake a "back of the envelope" quick determination of potential losses from a pre-designed seismic event. HAZUS also allows users to do rapid post-event assessments. The software was tested in September 2000 with the Labor Day earthquake in Yountville, California. The damage estimates predicted by HAZUS for an earthquake of the same size and magnitude as Yountville turned out to be very similar to the actual level of destruction and damage.

To be more useful at the local level for planning and mitigation activities, however, the default inventory of structures and infrastructure must be updated at the local level. Improvements in local risk factors (geology) are also required inputs for more detailed analyses at the local level. As is the case with many models, there are acknowledged limitations (FEMA 1997c). Among the cautions are:

- Accuracy of estimates is greater when applied to a class of buildings than when applied to specific buildings.
- The accuracy of estimated losses associated with lifelines is less accurate than those associated with the general building stock.
- There is a potential overestimation of losses, especially those located closed to the epicenter in the eastern United States due to conservative estimates of ground motion.
- The extent of landslide potential and damage has not been adequately tested.
- The indirect economic loss module is experimental and requires additional testing and calibration.
- There is uncertainty in the data and a lack of appropriate metadata (or sources) for many of the components.
- There is spatial error due to the reconciliation of physical models (e.g., ground shaking) with census-tract demographic data.

Financial Risks and Exposure: Private-Sector Approaches

The insurance industry has a long tradition of involvement in exposure assessment and risk management. Many of these tools and techniques are described in more detail elsewhere (Kunreuther and Roth 1998).

In most instances, financial risk assessment is based on probabilistic models of frequency and magnitude of events. In addition, proprietary data on insured properties, including the characteristics of those holdings are available and subsequently used to develop applications of expected to worst-case scenarios. For example, the Insurance Research Council (IRC) (IIPLR and IRC 1995) released a report on the financial exposure of states, based on the Hurricane Andrew experience. They found, along with increased population in the coastal area, that there is a significant rise in the value of insured properties, resulting in a potential loss of more than $3 trillion in 1993. This has increased since then.

Other examples of financial risk models that are used include Applied Insurance Research's catastrophe model and Risk Management

Solutions' Insurance and Investment Risk Assessment System (Heinz Center 2000a). To protect themselves from catastrophic losses, insurance companies utilize the reinsurance market as a hedge against future losses from natural hazards. Reinsurers use very sophisticated risk assessment models to calculate the premiums they charge to the insurance company and rely quite extensively on the scientific community (Malmquist and Murnane 1999). Reinsurers need to know the worst-case scenario (or probable maximum loss) and incorporate the following data in their calculations: hazard risk (intensity, frequency, and location), location of insured properties (homes, businesses, cars, etc.), existing insurance, and insurance conditions (deductibles, extent of coverage). Examples include PartnerRe's tropical cyclone model (Aller 1999) and Impact Forecasting's natural hazard loss estimation model for earthquakes and hurricanes (Schneider et al. 1999). The latter is especially interesting because it includes a mitigation application.

Ecological Risk Assessments

Most of the hazard loss estimation tools and techniques focus on the built environment such as buildings, transportation networks, and so forth. The loss of ecological systems and biodiversity is also important when thinking about vulnerability. These natural resources are the building blocks of modern economic systems. While there are many causes of ecological loss, most of them derive from human activities—urban encroachment, land-use changes, deforestation, introduction of foreign species, environmental degradation, and natural disasters. Historically, biodiversity loss was handled via individual species protection (endangered and threatened species) or through habitat conservation (mostly through national parks, wilderness areas, or biosphere reserves) (Cutter and Renwick 1999). One technique for the identification of these critical habitats and the species that live there is gap analysis, a program run by the National Biological Survey. Using remote-sensing and GIS techniques, gap analysis maps land cover, land ownership patterns, vegetation types, and wildlife to identify areas that are underrepresented in protected management systems, hence the name "gap" analysis (LaRoe et al. 1995, Jensen 2000). Gap analysis has been successfully used in locations throughout the nation to reduce ecological losses.

Another approach to ecological risk assessment is to examine the potential impact that human activities have on individual plants and animals or on ecosystems. Ecological risk is defined as the potential harm to an individual species or ecosystem due to contamination by hazardous substances including radiation. Ecological risk assessments (ERAs) provide a quantitative estimate of species or ecosystems damage due to this contamination by toxic chemicals. ERAs are similar in nature and scope to human health risk assessments, although the targeted species is different. The USEPA provides a set of guidelines for conducting ERAs, which have focused on single species, and single contaminants (USEPA 2000c). However, the science of ERAs is quickly evolving to include multiple scales, stressors (contaminants), and endpoints in order to achieve more robust and holistic risk assessments of threatened ecosystems, but it is not quite there yet (Suter 2000).

Comparative Risks and Vulnerability

Thus far, we have examined methods that help to estimate risk, assess vulnerability, and estimate financial and ecological losses from a single phenomenon. There are, however, techniques especially designed to provide a comparison among places with respect to risk levels and vulnerability, and these are described below.

Comparative Risks Assessment (CRA)

In 1987, the USEPA released a report, *Unfinished Business* (USEPA 1987), which set the stage for a realignment of agency priorities. There was a fundamental change at the agency—a movement away from pollution control to pollution prevention and risk reduction. To more effectively utilize resources and prioritize and target the most severe environmental problems, the USEPA developed the CRA tool (Davies 1996). Although many argue that the results of the ranking exercise are sensitive to the specific procedures used including categorization (Morgan et al. 2000), CRA does allow the incorporation of technical expertise, stakeholders, and policy makers in targeting activities and prioritizing risks and resources to reduce them. A number of case studies have been completed with varying degrees of success, the most important one being the tendency of the process to build consensus among the participants on the prioritizing of environmental problems (USEPA 2000d).

RADIUS

As part of the International Decade for Natural Disaster Reduction (IDNDR 1990-2000), the United Nations established an urban seismic risk initiative. The Risk Assessment Tools for Diagnosis of Urban Areas Against Seismic Disasters (RADIUS) developed practical tools for seismic risk reduction in nine case-study cities, especially in the developing world (Okazaki 1999). In addition to an assessment of seismic risk (exposure) and vulnerability, an action plan for preparedness against future earthquake disasters is also required. Based on input data (population, building types, ground types, and lifelines), and a hypothetical scenario earthquake such as the Hanshin-Awaji earthquake in Kobe, Japan, the output indicators include seismic intensity (Modified Mercalli Scale), building damage estimates, lifeline damage, and casualties. In addition to the nine detailed case studies, 103 other cities are carrying out seismic risk assessments based on this methodology (Geohazards International 2000). On a more general level, urban seismic risks can be compared using the Earthquake Disaster Risk Index (EDRI) (Cardona et al. 1999). The EDRI compares urban areas for the magnitude and nature of their seismic risk according to five factors: hazard, vulnerability, exposure, external context, and emergency response and recovery. Urban areas can be compared using any or all of these dimensions (Figures 2-5a and 2-5b).

Disaster-Proneness Index

The United Nations Disaster Relief Organization produced an assessment of natural hazard vulnerability at the beginning of the IDNDR in 1990. Using a 20-year history of natural hazards and their financial impact on individual nations (expressed as a percentage of GNP), the Disaster-Proneness Index provides a preliminary evaluation of the relative hazardousness of nations (UNEP 1993). Island nations, especially those in the Caribbean, rank among the top 10 on this index, as do a number of Central American (El Salvador, Nicaragua, Honduras), African (Ethiopia, Burkina Faso, Mauritania), and Asian (Bangladesh) countries. When mapped, the most disaster-prone countries are in the developing world. This is not surprising, given their biophysical and social vulnerability and the inability to respond effectively in the aftermath of the event or to mitigate future disasters (Cutter 1996b). There has been no subsequent update or revision of this original index.

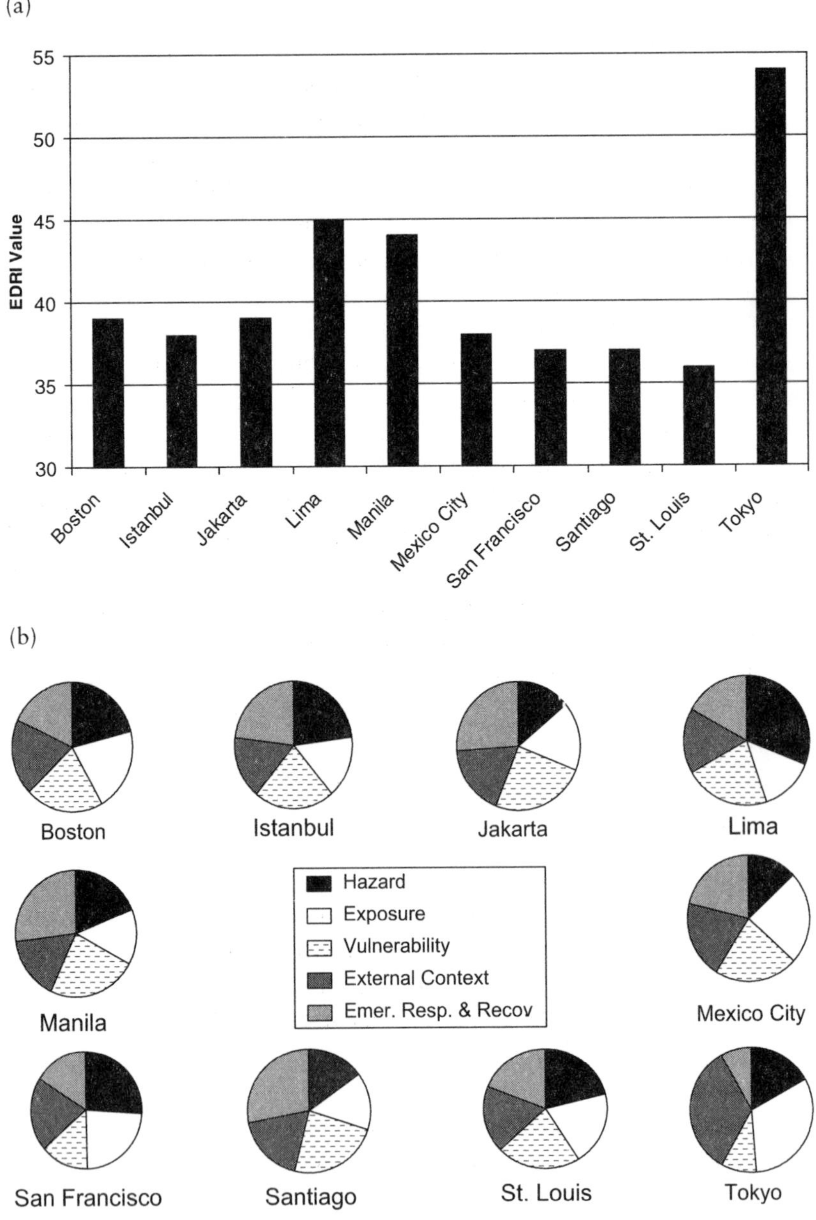

FIGURE 2-5 Earthquake disaster risk index: (a) comparison of 10 cities according to their overall earthquake disaster potential and (b) individual components of vulnerability that contribute to it. Source: Davidson (2000).

Pilot Environmental Sustainability Index

A newly emerging vulnerability index uses the concept of sustainability in understanding vulnerability to hazards. In January 2000, the World Economic Forum, the Yale Center for Environmental Law and Policy, and the Center for International Earth Science Information Network released the Pilot Environmental Sustainability Index. This prototype was designed to measure sustainability of national economies using one composite measure with five components: environmental systems (healthy land, water, air, and biodiversity), environmental stresses and risks, human vulnerability to environmental impacts, social and institutional capacity to respond to environmental challenges, and global stewardship (international cooperation, management of global commons). While still in its developmental stage, the Pilot Environmental Sustainability Index shows much promise in differentiating countries that are achieving environmental sustainability or are ranked in the top quintile (New Zealand or Canada) from those that rank in the lowest quintile (Mexico, India) (World Economic Forum 2000).

CONCLUSION

This chapter examined the range and diversity of current methods used to assess environmental risks and hazards and societal vulnerability to them. Some approaches are hazard specific whereas others incorporate multiple hazards. Global comparisons of risks and hazards exist, yet most of the assessment techniques and their applications are hazard specific or geographically restricted. In some instances the state-of-the-art science and technology have not made their way into practice, whereas in other cases the needs of the emergency management community are driving the development of new and innovative approaches to measuring, monitoring, and mapping hazard vulnerability.

Our review of the contemporary hazard and vulnerability methods and models did not include two important considerations that will drive the next generation of models and methodologies. First, we did not provide an explicit discussion of the underlying and contextual factors that increase the vulnerability of people and places: urbanization, demographic shifts, increasing wealth, increasing poverty, labor markets, cultural norms and practices, politics, business, and economics. Changes in any of these contextual factors will either amplify or attenuate vulnerability in the future or result in greater regional variability in hazardousness.

Second, we did not address the issue of how current hazards and risk assessment methods and practices might actually contribute to the relocation of risk and vulnerability (either geographically or into the future). This transfer of risk and vulnerability (variously called risk relocation or risk transference) (Etkin 1999) in either time (present to future) or space (one region or area to another) is an important consideration in assessing hazards and vulnerability. Yet our current methods and models are woefully inadequate in this regard, especially in determining some of the ethical and equity questions involved in the transference itself.

The implementation of public policies based on our hazards assessments have the potential to reduce vulnerability in the short term (a sea wall to protect against coastal erosion) yet they may ultimately increase vulnerability in the longer term, thus affecting the next generation of residents. Understanding, measuring, and modeling future risk and vulnerability are among the many environmental challenges we will face in the coming decade. The next chapter examines another future challenge—how to visually represent hazards and risks in ways that are both scientifically meaningful and understandable to a general audience.

CHAPTER 3

Mapping and the Spatial Analysis of Hazardscapes

Michael E. Hodgson and Susan L. Cutter

Maps are a fundamental means of communication. They have been used since prehistoric times to indicate directions to travelers, describe portions of the Earth's surface, or record boundaries including zones of danger. The collection, mapping, and analysis of geographic information are essential elements in understanding how we live with, respond to, and mitigate against hazards. We begin this chapter with a primer on the fundamental concepts in mapping—scale, resolution of geographic data, and characteristics of spatial databases. We then provide a brief history of mapping generally, and hazards mapping specifically. The chapter concludes with a discussion on the role of technological advances in influencing the mapping and the spatial analysis of societal response to hazards.

INFORMATIONAL NEEDS AND INPUTS

The data we use in hazards assessment and response have two fundamental characteristics: a time (or temporal) dimension and a geographic (or spatial) dimension. The requirements for each may be quite different, depending on the application. For example, the data needed for post-event emergency response (rescue and relief) are quite different from the information we need for longer-term recovery and mitigation efforts. Similarly, the geographic extent of our data varies depending on the nature of the event.

Temporal Considerations

In hazards applications, we are concerned with when the data were collected, the time interval required for the collection, the lag time between when the data were collected and when we can use them, and finally, how frequently new data are collected. For example, censuses of population and housing at the block level are extremely important for modeling population at risk from future flooding. Yet, these population surveys are made once every 10 years (frequency). Even after the population census is conducted, it may be 2 years or more before the data are made public (lag time). So, communities often use population projection data (for updates) or simply use "old" data because they cannot afford to collect such extensive demographic data on their own. Clearly, these time-dependent characteristics influence the types of questions we can answer and the ways in which data can be used.

Often, special data collection efforts are required in order to conduct hazards assessments that include the most current information possible. The following provides a good example. The collection of aerial photography for post-disaster emergency response or damage assessment (e.g., after a hurricane) can be scheduled upon demand, such as the day after the event. Because it is a snapshot, the photography represents the landscape on that day (collection date). Damage assessments from aerial photography typically are conducted by comparing post-disaster to pre-event photography. Unfortunately, the last pre-event photography may be quite old (e.g., 5 to 10 years). An example is provided by the pre- and post-damage assessments from Hurricane Andrew. The State of Florida was fortunate to have recent (within 1 year) pre-disaster imagery (Figure 3-1a, b), which made pre- and post-event comparisons easier. The updating of flood insurance rate maps (FIRMs) provides another example of the temporal dimension in hazards mapping, where the time interval and frequency of revisions may pose pre-impact planning problems especially for rapidly urbanizing areas. In some places, FIRMs may be more than 10 years old and may not reflect or may inadequately represent the recent growth and development in that community.

(a)

(b)

FIGURE 3-1 Mobile home park in Homestead, Florida (a) before and (b) after Hurricane Andrew in 1992. Source: Photography scanned by M. E. Hodgson from a pre-event photo from NAPP USGS, and post-event photography flown by Continental Aerial Surveys.

Spatial Scale, Resolution, and Extent

Just as geographic data have temporal dimensions associated with their collection or analysis, they also have spatial characteristics. One of the first decisions a cartographer makes is to determine how much of the Earth's surface is to be represented on the map. This is known as map scale; it is the relationship between the length of a feature on a map and the length of the actual feature on the Earth. Small-scale maps (such as a map of the world) cover a large proportion of the Earth's surface, but offer very little detail about it other than broad generalizations about patterns. Larger-scale maps are just the opposite—they portray a smaller area but with greater detail. An example would be a city road map. Large-scale and small-scale maps each have their own purpose and use in hazards. We would not presume to navigate the city with a wall map of the world, nor would we want to use the wall map to tell us how to get to the Federal Emergency Management Agency (FEMA) headquarters in Washington, D.C. Thus, knowledge of scale is an important determinant of what type of geographic data should be collected and the ultimate purpose of the map.

All geographic data should include a description of the spatial characteristics—observation unit (or spatial resolution), collection/reporting unit, and spatial extent. Scientists from different disciplines often incorrectly use the term "spatial scale" to refer collectively to one or more of these characteristics. For example, suppose we wanted to predict the geographic distribution of anticipated damage associated with a future hurricane for all major metropolitan areas in the United States. To do this, we would construct a building loss-wind speed functional statistical relationship based on a sample of individual homes in a coastal area. Next, we would forecast the spatial distribution of total future losses, using building characteristics within census tracts. At what scale are we operating? It might be most appropriate to say that the damage-wind speed estimate is at the individual building scale (our observational unit) but the forecast of future losses is at the census tract scale (our collection/reporting unit). Even though the spatial extent of our study was nationwide (at least all metropolitan areas in the country), it would be inappropriate to say we had conducted a national-scale study. As this example illustrates, the use of the term spatial scale is often so contextual and audience-specific that it is confusing to others when we say our study is at such-and-such a spatial scale. Despite this confusion, spatial scale is still an important parameter in geographic data collection and data storage.

Selection of appropriate spatial characteristics is critical for hazards and vulnerability assessments (Been 1995, Cutter et al. 1996). It is arguably one of the primary reasons why different relationships are observed between risk variables in diverse analyses. Research in nonhazards fields, for example, has documented that the correlation between variables (e.g., percentage minority population and mean family income) varies for the same spatial extent as the size of the observational unit changes (Clark and Avery 1976). In general, the correlation between variables increases as the size of the sampling unit increases (moving from block groups to counties, counties to states, sampling plots to ecosystems). In geographic terms, this is called the modified areal unit problem or, more commonly, the "ecological fallacy" (Cutter et al. 1996).

So, what is the appropriate spatial resolution of analysis for hazards and vulnerability assessments and hazards mapping? The immediate response is that it depends on the purpose of the assessment or map and the availability of data. Ideally, we would want to achieve a balance between the best resolution possible (the greatest level of detail, such as the smallest census unit for geo-referenced population data), our understanding of the processes involved (based on individuals or groups), and our ability to process the data. From a hazards standpoint, the appropriate resolution of analysis must be commensurate with the ability to delineate the risk or hazard studied. For example, it makes no sense to use county- or census-tract-level data for estimating the characteristics (e.g., race, income) of the population at risk around a noxious facility—if the risk zone is only one-quarter mile. The spatial differences between the small, one-quarter-mile radius area at risk and the larger census tracts (or much larger counties) is too great. This example is analogous to estimating the kinds of fish along the Toledo shoreline based on the varieties and densities of fish in all of Lake Erie.

Fundamental population characteristics (e.g., age and race) are published in the Decennial Census at the block level; more detailed characteristics are available only at block-group and larger census units. Thus, some hazard impact analyses may be conducted at block-level scales (potential value of the housing stock) whereas others may be limited by the available data and only focus on census tracts or counties (social vulnerability based on personal income). Furthermore, consider the ominous task of analyzing the U.S. population at risk from some hazard at the block level. Processing the spatial data for over seven million blocks (the number of blocks in the 1990 Decennial Census) would put severe burdens on most computers even if the models were simplified.

Graphical Representations

Maps are used to depict the spatial relationships of objects. There are many different types of maps, but the most common are classified as thematic maps because they concentrate on the spatial variability of *one* phenomenon, such as rainfall or tornado touchdowns. There are a number of specific cartographic representations that provide us with additional information about the phenomena we are mapping (Figure 3-2).

Choropleth maps (the most commonly used in hazards research) show the relative magnitude of the variable using different shadings or colors. In this way, a reader can easily determine those areas that have more rainfall (or tornadoes) by simply looking at the darkly shaded areas on the map (see Chapter 6). Contour maps are another example. Contour maps represent quantities by lines of equal value (isolines) and thus illustrate gradients between the phenomena being mapped. We are most familiar with them for depicting elevation changes on topographic maps, but isoline maps also are used to display regions of high and low pressure on weather maps.

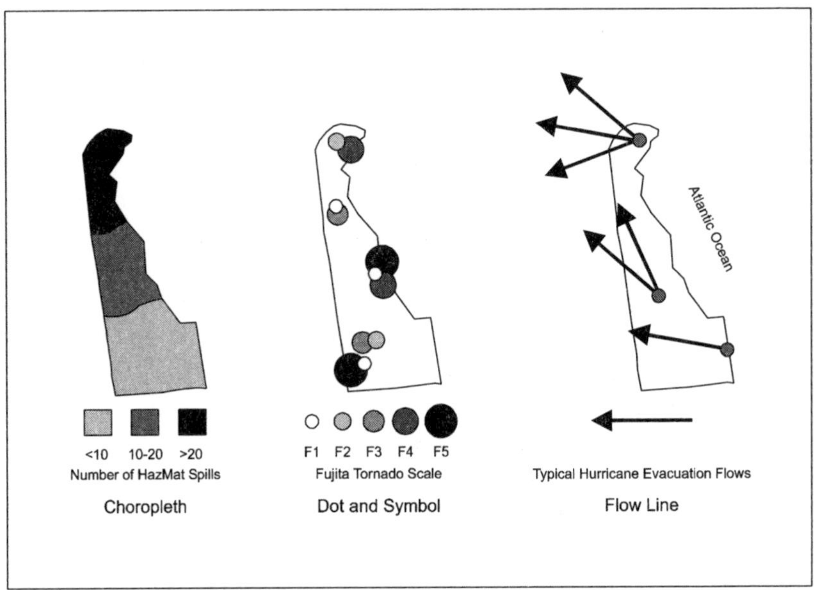

FIGURE 3-2 Different cartographic representations of hazards data.

Dot maps show the location and distribution of specific events, such as tornadoes. A variant on the dot map is the symbol map, which uses different sizes or shapes to indicate a quantity at a specific location. For example, different-size circles could be used to illustrate the different intensities of tornadoes based on the Fujita Scale. Hazards information is also represented by line maps, which illustrate actual or potential movement or flows. The volume of hazardous waste transported from one location to another, for example, could be depicted using such a map. Similarly, the number of evacuees leaving specific coastal towns could be represented that way. Finally, animated maps can be used to graphically illustrate changes in phenomena over time, such as the population of the United States, the growth of a city, or the likely storm surge from an impending hurricane (Collins 1998).

IMPROVEMENTS IN DATA COVERAGE AND ACCURACY

During the past 25 years, our knowledge about risk, hazards, and vulnerability has improved. Over the same time period, we have seen major advances in techniques for collecting hazards data with increased accuracy. Better data result in improved hazard and vulnerability assessments.

Monitoring and Surveillance

The greatest advances in technology have been in the areas of monitoring and surveillance of hazards. The mid-1970s saw the implementation of real-time detection of seismic events via digital monitoring networks (Stewart 1977). As the 1980s progressed, researchers installed denser networks, trying to increase the global coverage of sampling points. For example, by the mid-1990s, over 46,000 seismometers had been deployed, but in a very uneven spatial pattern (Wysession 1996). Although we can now monitor global seismic activity, there are still some regions about which we know relatively little (seismically) because of the lack of monitoring at the subglobal level.

Another hazard that has been the focus of increased surveillance and monitoring over the past 25 years is severe meteorological storms resulting from extratropical cyclones and convection. The development of WSR-88D Doppler radar has increased forecasters' ability to predict and track tornadoes, hail, and wind. Further refinements in the technology have enabled forecasters to examine three-dimensional data to look for the distinct "hook" or signature of a forming tornado (see Chapter 2). In addition, Doppler radar has been used to estimate precipitation at given locations in order to predict possible flood hazards (Figure 3-3, see color plate following page 22). Doppler radar systems now are installed in most urban areas (airports) across the country as part of the National Weather Service's (NWS) modernization program. However, there are many remote areas where Doppler radar still is unavailable.

Perhaps the greatest technological advances in hazards surveillance and monitoring are remote-sensing technologies. In 1960, the National Aeronautics and Space Administration (NASA) launched the first experimental weather satellite, TIROS 1. Because of the success of the polar-orbiting TIROS and NIMBUS (1964-1974) programs, National Oceanic and Atmospheric Administration (NOAA) also began the Geostationary Observational Environmental Satellite (GOES) program in 1975. Versions of these satellites continue to operate today, providing scientists with access to near real-time images of the weather around the globe. The ability to view severe storms and hurricanes using more than the visible bands of the electromagnetic spectrum has allowed meteorologists to examine the tracks and internal characteristics of storms, which was not possible two decades earlier (Lillisand and Kiefer 1994). It has also assisted in the mapping of the impact areas of natural hazards such as lava flows, natural oil seeps, floods, droughts, and dust storms (Abrams et al. 1991, Simpson 2000).

In addition to meteorological hazards, remote sensing has been used to monitor wildfires (Figure 3-4, see color plate following page 22) and air pollution. The U.S. Geological Survey (USGS) and NASA initiated the Earth Resource Technology Satellite (ERTS) program, resulting in the deployment of seven satellites to collect Earth resources information. The first of these, ERTS-1, was launched in 1972 and had a spatial resolution of 79 by 79 meters. LANDSAT 1, (the renamed ERTS satellite) became operational in 1978 and subsequent satellites were launched throughout the next two decades, tracking small-area events with less expense and great efficiency, and providing a potential wealth of data for the hazards community.

Satellite data also have been used to assess post-event damages and thus aid in resource allocation during the hazard recovery process. After Hurricane Andrew, for example, remote-sensing technologies helped to quickly determine the extent of damage for the disaster declaration (Davis 1993). In the aftermath of Hurricane Hugo, remote-sensing data were used to determine the nature and extent of damaged forest resources in coastal South Carolina (Cablk et al. 1994). Previous property damage assessments were restricted to either field reconnaissance, which took weeks or months after the event for ground crews to enter the affected area, or through aerial and digital-frame photography, but satellites could not provide enough detailed information to be helpful. Thus, the spatial and temporal resolution of sensor systems has limited the direct application of remote sensing to hazard studies up to now.

The newest generation of satellites (releasing unclassified data) has spatial resolutions on the order of 1 by 1 meter, enabling even more precise monitoring of hazards (and their impacts) at the local level (Jensen 1996, 2000). For example, Space Imaging Inc.'s IKONOS platform includes multispectral digital-frame cameras that can "sense" infrastructure such as roads and pipelines and even individual homes. This high-resolution pre- and post-disaster imagery can provide a valuable source of information during disaster response and recovery (Cowen and Jensen 1998, Jensen and Cowen 1999). In 1999, the TERRA satellite was launched, ushering in a whole new generation of remote-sensing satellites to assist in hazard surveillance (King and Herring 2000). Initially geared toward monitoring of global climate change indicators, a number of the sensors on the TERRA satellite have tremendous utility for hazard assessments (Table 3-1). The potential of the newer technology is only now beginning to be realized.

Finally, as part of its Map Modernization Plan, FEMA is currently assessing the applicability of remote-sensing techniques in producing National Flood Insurance Rate maps (FEMA 2000b). Specifically, FEMA is experimenting with the use of LIDAR (light airborne detection and ranging) and IFSAR (Interferrometric Synthetic Aperture Radar) sensor systems to generate digital elevation models and digital terrain maps that could be used to project flood-prone areas based on elevation. The experiment is designed to see if the application of this technology can predict the geographic extent of flooding (e.g., 500-year flood event or 100-year flood event) associated with Hurricane Floyd in Tarboro, North Carolina (J. R. Jensen 2001, personal communication).

The spatial and temporal characteristics of data and the subsequent maps produced from them will vary with the needs of the researcher or emergency manager. In mapping individual structures in hazard or risk assessments, the spatial resolution must approach the 1 by 1 meter pixel size or better (Welch 1982, Cowen and Jensen 1998). The rapid assessment of post-disaster damage for emergency response requires a spatial resolution of 0.25 to 1 meter. However, this information must be available to the emergency manager within 1 to 2 days after the event (Cinti 1994, Wagman 1997) to be of greatest use. On the other hand, pre-impact planning and mitigation often requires less spatial resolution (less

TABLE 3-1 TERRA Satellite Sensors Used to Monitor Hazards

Sensor	Spatial Resolution	Attribute	Targeted Hazard
Aster	15-90 m	Can target selected sites or areas	Floods, wildfires, volcanic eruptions, earthquakes
CERES	20 km	Monitors radiation fluxes	El Niño, La Niña
MOPITT	22 km	Monitors trace pollutants and their sources	Pollution, chemical releases
MISR	275 m-1.1 km	Monitors clouds and smoke plumes	Smoke plumes from industry, volcanic eruptions, etc.
MODIS	250 m-1 km	Covers entire planetary surface every 2 days	Volcanic eruptions, floods, severe storms, drought, wildfires, snow cover

Sources: King and Herring 2000, Jensen 2000.

detailed information) and a longer time frame (months to years). Both need to be considered when using remotely-sensed data to map hazards.

Data and Improved Collection Accuracy

Historically, the collection of data for hazards mapping began with fieldwork. The seminal work by John Snow on cholera used individual observations of cholera cases—where the victims lived in the city—to find the source of the health threat (Frerich 2000). Fire insurance maps during the 1800s also used field observations to note the location and construction materials of every building in the urbanized area. Even today, the "windshield survey" is the fundamental method for building a spatial database of damaged homes after a hurricane event.

Advances in field survey methods for building spatial databases include the mobility of the observer and improved methods for determining geographic position (Goodchild and Gopal 1992). Largely developed in the 1980s, the Global Positioning System (GPS) provides a very accurate method for determining the geographic position of the observer. Compared to other methods of locating positions, GPS is essentially an all-weather, 24-hour/7-day global system that is relatively cost-free to the user, who needs to purchase only the receiving device. Single-point positions are accurate to within 30 meters with the inexpensive handheld receiver (costing about $100). This new, inexpensive, and accurate method of determining geographic position is used widely in emergency response and in transportation-related hazard studies. Documentation of damage at a specific location from a hazard event has been greatly improved. Field personnel merely have to push one button on the automobile-mounted GPS to locate and record the position of the noted damage. Another example is monitoring the movement of hazardous materials using GPS in a truck, barge, or train.

THE SCIENCE AND ART OF MAPPING

Rapid technological advancements during the twentieth century (photography, aircraft) enabled us to expand not only the coverage of mapped areas of the Earth, but the efficiency with which it was done. We no longer needed to complete field surveys for mapping, but could use fixed-wing aircraft that took photographs (which were then converted into maps). This method had the advantage over field surveys in that larger areas could be covered in more detail, resulting in both time and labor savings. In February 2000, the ultimate mapping project took place when the Space Shuttle *Endeavor* mapped 80 percent of the Earth's surface in 10 days, creating three-dimensional topographic maps with 30-meter resolution (accurate within ± 30 meters of their true location) (Chien 2000).

The science of mapping—cartography—has a long and rich history and tradition. The oldest known map (on a clay tablet) was found in Mesopotamia and is thought to be a city map of landholdings (Robinson et al. 1978). Now, such maps are called *cadastral* (or tax assessment) maps. Starting with the ancient Greeks and Romans—and later Ptolemy (known as the father of cartography) in ancient Egypt and Phei Hsiu in ancient China—maps were a standard form of communicating geographic information about places. The quality and accuracy of many of these early maps were unchallenged for more than 10 centuries! The art of cartography was lost during the Dark Ages in Europe, but was reintroduced by the Arab cartographer, Idrisi. In addition to the increased curiosity about faraway places, seemingly accurate sailing charts (the forerunner of contemporary nautical charts) appeared, both of which stimulated additional interest in map making.

The rapid advancements in map making during the Renaissance were largely due to the great voyages of discovery and the development of the printing press, which allowed multiple copies of maps to be reproduced. Modern cartography dates to the seventeenth century with the use of a Cartesian grid system to more accurately position and orient the map. National surveys, especially in France and England, delineated not only their territory, but also the physical and human features within it ushering in what we now know as topographic mapping. Thematic mapping (maps of specialized subjects such as climate, weather, population density, landforms, and even hazards) became the norm during the nineteenth century when the process of lithography made those maps easier and less expensive to duplicate, and thus more widely distributed than ever before.

The development of the computer has enhanced cartography more than any other single technological advancement during the twentieth century. The integration of geographic information along with environmental and socioeconomic data has resulted in the development of the Geographic Information System (GIS), the dominant force in today's mapping of hazards and their impacts.

Hazards Cartography

It might be argued that the work of John Snow, a physician in London, England, was the earliest example of applying cartographic approaches to hazards—in that case epidemiology (Monmonier 1997, Timmreck 1998, Frerich 2000). Snow recorded cholera observations in the mid-1840s and then created spatial and temporal distributions of the infections and deaths in a small area of London. By comparing the distribution of cholera cases to water supplies, he was able to suggest hypotheses about the origin (i.e., the water supply), transmission, and control of the disease. After mapping and analyzing the spread of cholera, he finally removed the pump handle from the contaminated water source on Broad Street to control future cholera outbreaks (Frerich 2000).

The development of weather maps was another early example of the use of cartography to convey hazards information. First suggested in 1816 by a German physicist, Heinrich Brandes, daily weather maps did not routinely appear until the 1860s in Europe and almost a decade later in the United States (Monmonier 1999). Another early example of hazards cartography was the mapping of urban structures for fire insurance purposes. Fire insurance maps date from the late eighteenth century, with the earliest known map (of Charleston, South Carolina) produced in 1790. The dominant company producing such maps, the Sanborn Company, began creating fire insurance maps in 1870. Field research detailing each structure's position, construction type, and purpose was used to make each Sanborn map. These fire insurance maps were completed for most urban areas in the United States and updated periodically until the mid-1900s. Although Sanborn maps are no longer produced, the Sanborn Company still exists and sells copies of the older maps. Today, the primary use of the now-dated Sanborn maps is for researching prior use of sites to determine whether any hazardous materials storage or manufacturing was present at any time in the past. Some states require the use of historic materials such as fire insurance maps in design studies and local zoning approvals for redeveloping industrial sites in urban areas, called "brownfields development."

The systematic "mapping" of human occupance of hazardous areas began with Gilbert White and his students at the University of Chicago during the 1960s and 1970s. Aimed at understanding human responses to extreme natural events, a significant portion of the comparative research both within the United States and internationally involved the estimation and delineation of the human habitation of areas subject to extreme natural events (White 1974, Burton et al. 1993). These field observations linked the increasing level of losses to a number of factors, including increasing human excursions into risky areas. A comparison of the disproportionate effect of pollution on various social groups using a sample of American metropolitan areas was another early cartographic view of risks and hazards (Berry 1977).

Perhaps the most widely known hazard maps are those done by FEMA's Flood Hazard Mapping Program. Begun in 1968 under the auspices of the National Flood Insurance Program (NFIP), floodplain maps (and the subsequent FIRMs) have been completed for more than 20,000 communities nationwide (Platt 1999). Over 100,000 map panels (approximately the size and scale of the USGS 7½-ft quadrangle) were originally produced in this program (Figure 3-5). The original intent of these maps was for lenders, property owners, and real estate agents to determine if properties generally were located in areas of high probability of future flooding (designated 100-year and 500-year floodplains).

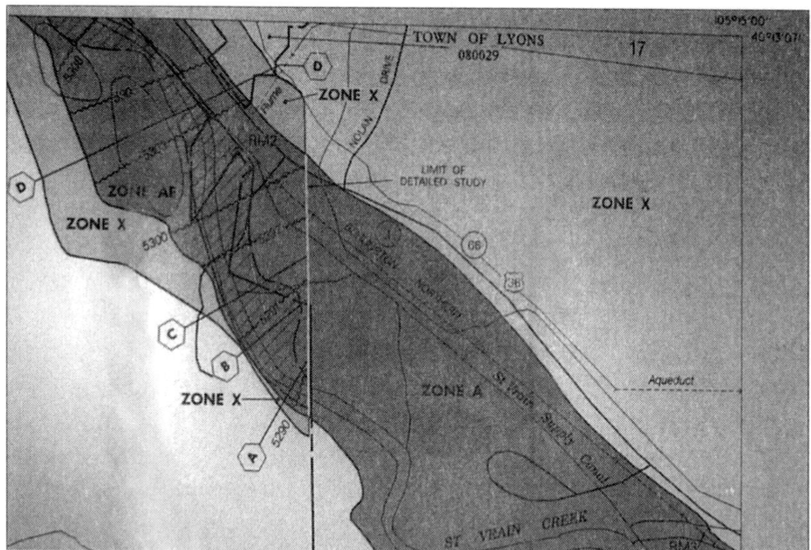

FIGURE 3-5 Example of a FIRM showing the locations of the 100-year (A zones in dark gray) and 500-year (X zones in light gray) flood zones. Source: FEMA.

Moving from Manual to Automated Hazards Cartography

The creation of maps by manual methods has long passed. Even the use of strictly automated cartographic applications for hazards work has been replaced by the use of GIS. The unique capabilities of GIS to combine other geographic information distributions for analysis with the specific hazard (e.g., disease, fire) and then to simply map the spatial distributions have made it the technology of choice. Thus, one seldom discusses cartographic processes today without GIS.

There are many examples of the use of GIS to produce hazard maps. Goldman (1991) produced an atlas of toxins and health that mapped the distribution of industrial toxins, pollution, and health indicators by county for the entire United States. Aptly titled—*The Truth About Where You Live*—this atlas provides a striking picture of hazard zones throughout the country. FEMA's (1997a) multihazard assessment is another example, illustrating the geographic variability in individual hazard occurrences throughout the country. More recently, an atlas of environmental risks and hazards has been produced for South Carolina (on CD-ROM), illustrating the geographic patterns of hazards and their consequences by county within the state (Cutter et al. 1999). The natural hazards map of North America produced by the National Geographic Society (Parfit 1998, Tucker 1998) provides a generalized depiction of the primary hazards affecting the continent. Despite the necessary level of generalization, the map provides extensive coverage of historic hazard events at this scale. These are but a few of the many examples of hazard mapping that are occurring today.

SPATIAL ANALYSIS AND THE GIS

Spatial analysis is a term used to describe a set of tools that examine patterns in the distribution of human activity or environmental processes or both as well as movement across the Earth's surface. There are many tools for conducting spatial analyses: statistics, mathematical models, cartography, and GIS. Two are particularly important for hazards: the GIS and its refined cousin—the spatial decision support system (SDSS). Both provide the foundation for mapping, analyzing, and predicting hazards and impacts. Evolution of the GIS has had a profound impact on hazards research and application, especially in studying hazards in a spatial context.

GIS

The development of remote sensing, the creation of commercial GIS software, and the proliferation of digital geographic databases were key to the widespread use of GIS. The initial development of computers was limited to federal agencies or a few universities during the 1950s and 1960s. Beginning in the late 1970s, and particularly in the early 1980s with the creation of relatively powerful yet inexpensive personal computers, GIS software was developed and marketed by private industries. Leaders in GIS software development (e.g., ESRI, Intergraph, ERDAS, and MapInfo) are improving the applications to all aspects of emergency management, from preparedness to response to recovery (Dangermond 1991, Johnson 1992, Newsome and Mitrani 1993, Beroggi and Wallace 1995, Marcello 1995, Carrara and Guzzetti 1996, Radke et al. 2000).

Today, the GIS is defined as a computer-based method for collecting, storing, managing, analyzing, and displaying geographic information. Geographic information (locational and attribute data) is collected with a GIS by either using a GPS receiver or conventional survey methods in the field or by conversion of remotely sensed imagery or existing maps into digital form. Alternative methods for acquiring digital geographic information include data purchases from private companies or data acquisition from state or local agencies. Also, geographic data can be obtained at minimal or no cost through federal agencies, such as the USGS, NASA, or NOAA.

Modern GISs contain a rich set of tools for analyzing geographic information. Fundamentally, a GIS allows diverse sets of geographic data to be put together in overlays so that the relationships between the different data "layers" can be analyzed. An example of a simple overlay application would be to use a digital map of residential houses and a digital map of the 100-year flood zones to determine the homes within the flood zone. By summing the value of structures within this high-risk area, an estimate of economic vulnerability could be made (Figure 3-6). The biophysical nature of the hazard (e.g., flood, wind, or ground failure) can be modeled by GIS applications as well (Carrara et al. 1991, Zack and Minnich 1991, Chou 1992, Mejia-Navarro and Wohl 1994). The geographic distribution of such hazard probabilities and/or magnitudes (potential ground motion with earthquakes) is then overlain on the distribution of human settlement patterns. Fundamental work with earthquake hazards was begun in this area by the Association of Bay Area Governments and has been extended to the entire United States at the

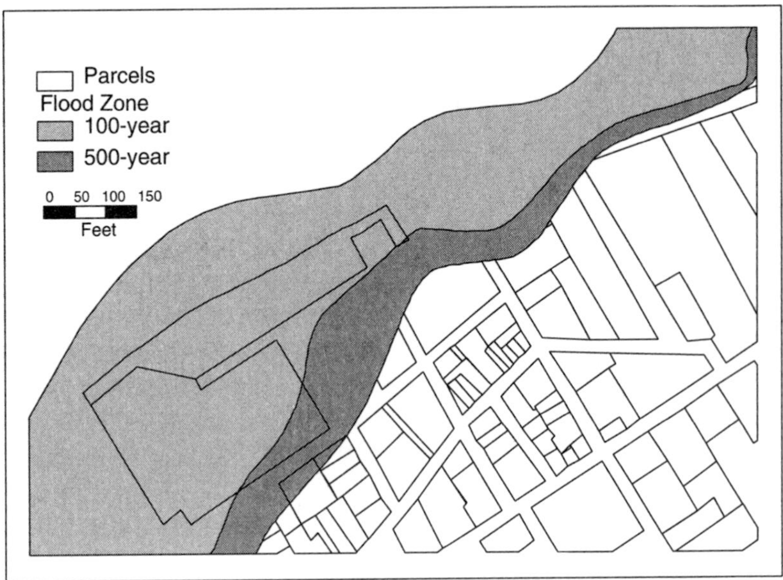

FIGURE 3-6 Intersection of flood zones and land parcels in Snow Hill, Maryland.

census-tract- and zip code-level scales of analysis with FEMA's loss estimation tool, HAZUS (see Chapter 2).

The GIS also has been used to assess hazardous waste transport and management (Estes et al. 1987, Stewart et al. 1993, Brainard et al. 1996). On the social side, the GIS has been used to identify environmental injustice based on toxic risks (Chakraborty and Armstrong 1997, Cutter et al. 2001) and in evacuation planning (Cova and Church 1997). The use of GIS analysis is not limited to aggregate data, such as census units, nor is it limited to published sources of information. Palm and Hodgson (1992a, b), for example, combined survey research and GIS methodologies to study individual and aggregate attitudes and behavior toward seismic risks.

SDSS

An SDSS is designed to address a specific problem for which numerous criteria are used to determine the selected course of action or policy alternative. Information included in the system guides the users or stakeholders through the process of implementing different scenarios. Thus, the SDSS supports decision makers by producing several alternative outcomes (such as variable siting patterns) based on different criteria and weights (or levels of importance) for the criteria. A good example of an SDSS for planning is the USEPA's LandView III. That system is used in evaluating inequitable patterns of hazardous facility siting (USEPA 2000e). It contains geographic background data (roads, streams, etc., from the Census TIGER/Line database), jurisdictional boundaries, USEPA databases on toxic substances and releases, and demographic data from the U.S. Decennial Census. Although the system allows the creation of maps and other cartographic products, another use of the system is to estimate the number and characteristics of people within a specific radius of a hazard site.

The SDSS has been used in emergency management for a number of years and has great potential for sustainable mitigation of hazards (Mileti 1999). With improved computer capabilities, the potential integration of disparate information is now possible. Decision support systems can aid in the allocation of resources, in developing scenario-based training exercises, and most importantly, in disaster management and incident control. For some applications, the development of GIS software has reached such a level of maturity that it is increasingly integrated into emergency management and response. The digital atlas of Central America in the aftermath of Hurricane Mitch by the USGS Center for Integration of Natural Disaster Information (CINDI) is one example (USGS 1999). These data, provided in map form and in GIS digital layers, served as a crucial resource for establishing critical needs and subsequent resource allocation for short-term relief efforts.

Unfortunately, the full functionality of a GIS is still limited to experienced users in each application field, be it floodplain management, water quality modeling, or seismic mapping. Less sophisticated users rely on pre-configured "GIS-based" packages, such as HAZUS, which have extremely limited decision-making options imbedded within them. Consider the following scenario: In a potentially hazardous situation such as an approaching wildfire or residences near a hazardous materials spill, it would be more desirable to systematically notify residents within the specific wildfire path or zone of danger via telephone rather than by a more "generic" and nontargeted approach such as a siren. This is already being done with a reverse Emergency-911 (E-911) system, where the GIS is used to identify risk areas and then notifications are issued. Such notification systems are already in use, for example, along Colorado's Front Range for wildfires.

A more sophisticated use is illustrated by the following example: In a hazardous material spill, an SDSS would need data on the location of the chemical spill, type of chemical, and release amount. Using a mathematical dispersal model, a digital topographic map, and existing meteorological conditions, a prediction of the dispersal pattern would then be modeled. The map of dispersion would be overlain on a map of residences to determine specific families at risk. The database of homes in the risk zone would be linked to a digital telephone directory and all homes would be automatically telephoned with a pre-recorded message. This entire process would take only a few minutes and involve an integration of data collection, GIS analysis, air diffusion modeling, and reverse E-911 phoning.

There are other GIS-based decision support systems used in public policy decision making (see Chapter 2), but the widespread implementation of SDSS is still not under way within the hazards community. It currently takes a specialist to manipulate this expert system, which provides an important impediment to its widespread use at this time.

DISTRIBUTING GEOGRAPHIC INFORMATION

There are many ways in which geographic information is distributed. Conventional methods include paper maps or digital databases. With new advances in technology, wireless communications and the Internet will become the most important ways to communicate hazards data in the future.

Wireless

Wireless communications will enhance hazards management in the future. For example, GPS technology has permitted the accurate determination of geographic positions (longitude and latitude) since the late 1980s, but its rapid declassification and commercialization opened opportunities for increased use in hazards management. Consider how this technology is now used to remotely monitor a hazardous materials shipment. The driver of the truck knows where the truck is at any point in time, but how does the command center know its location? The most widely used method of tracking trucks in the United States is with the Qualcomm OmniTRACS system, a type of automatic vehicle locator, which predates even the GPS. Although the positioning capability of the OmniTRACS was only about 300 meters, this level of accuracy was suf-

FIGURE 3-7 Wireless tracking of hazardous materials trucks. Typical location of Qualcomm transmitter on an 18-wheeler. The unit is the white dome located on top of the cab wind shell. Photograph by M. E. Hodgson.

ficient for interstate truck tracking. The key feature in the OmniTRACS is a transmitter/receiver based on a geostationary satellite (Figure 3-7). This system automatically polls the location of the moving truck (or any vehicle) every 15 minutes and the location is transmitted from the truck to a command center via the satellite. The OmniTRACS is the one employed by the U.S. Department of Energy for real-time tracking of their hazardous shipments. With the enhanced GPS capacity, this tracking system can be used to pinpoint transportation accidents involving the vehicle to within 100 meters (or 30 meters using a GPS) and thus enhances local response capability through better information on location, amount, and nature of the material spilled.

A more common wireless infrastructure, developed during the past 20 years, is the cellular telephone network. The combined analog and digital cellular network allows a more rapid rate of transmission (as compared to OmniTRACS). There is also a move toward incorporating GIS and GPS devices into cellphones in the near future. Unfortunately, even with the cellular telephone network today (over 75,000 cellular sites in the United States alone), geographic coverage of the United States is not complete. The dissemination of data over cellular networks is possible within most urban areas and major transportation arteries. Although cellular communication provides a much higher bandwidth than OmniTRACS, and thus faster transmission of data, it is not nearly as fast as through conventional landlines. Thus, hazards researchers and responders in the field may find it more useful to request specific data to be transferred directly to a laptop computer or, more commonly, a personal digital assistant.

Internet

The Internet has truly revolutionized the dissemination of information and knowledge and made it more accessible to everyone. More and more data are migrating to the Internet from both researchers and governmental agencies, and are becoming available with just a click of the mouse. In many ways, this has democratized information and enabled stakeholders to become more informed if they so desire. It has allowed local communities to become more knowledgeable about their own environmental vulnerabilities. However, there is just as much "junk" on the Internet as with printed information, and so, users must exercise caution when utilizing some of the information for hazards analysis and management.

Every day brings new advances in Internet applications and it is often difficult to keep up with favorite Web sites. Within the past few years, however, a number of important innovations have occurred, including the mapping of hazards on the Internet. Some of these are interactive, that is, you create or customize your own maps, whereas others are more static products. A number of these are discussed below.

USGS (www.usgs.gov)

The USGS hazards program includes earthquakes, volcanoes, floods, landslides, coastal storms, wildfires, and disease outbreaks in wildlife populations. In addition to fact sheets on each of these hazards and their management, the USGS also is involved in monitoring, predicting, forecasting, and mapping natural hazard events. For example, the geographic distribution of hazardous regions mapped for the contiguous United States is provided for earthquakes, volcanic hazards, landslide areas, major flooding, hurricane activity, and tornado activity in addition to a composite map of all of them.

Streamflow conditions are mapped daily for the entire United States (*http://water.usgs.gov/dwc*), providing real-time information on potential flooding and drought conditions. Links are also provided to drought monitoring and river conditions maintained by NOAA, the U.S. Department of Agriculture, the National Drought Mitigation Center, and the National Weather Service (NWS).

Another important mapping project by the USGS is the national seismic hazard mapping project prepared under the National Earthquake Hazards Reduction Program (NEHRP). These probabilistic maps are available digitally, as GIS coverages, and in hard-copy form (*http://geohazards.cr.usgs.gov/eq/html/genmap.shtml*). Covering the entire United States and for selected seismically active regions (California/Nevada, Alaska, Hawaii, central/eastern United States), these ground motion maps include peak ground acceleration. In addition to static maps, customized inquiries using zip codes or latitude and longitude can determine ground motion hazard values for specific locations for 10, 5, and 2 percent probabilities of exceedence in the next 50 years. These probabilities correspond to ground motions with return periods of 500, 1,000, and 2,500 years, respectively.

A variant of these static maps is a near-real-time map of earthquake intensities. In 1997, Southern California developed the TriNet, a digital seismic network that produces real-time "shakemaps" that include the location of the earthquake and the severity of ground-motion shaking (FEMA/NIST/NSF/USGS 1999). Information on monitoring-station coverage and events is available on the Internet (*http://www.trinet.org*) as are community-based felt-intensity maps (*http://pasadena.wr.usgs.gov/ciim.html*).

FEMA/ESRI On-line Hazard Maps (www.esri.com/hazards/index.html)

Through the Project Impact program, FEMA and Environmental Systems Research Institute, Inc. (ESRI) have developed an interactive multihazard mapping program. The interactive nature of the mapping is a significant improvement on existing Internet mapping applications. Although limited to historical occurrences of selected hazards (earthquakes, hail, hurricanes, tornadoes, windstorms) as well as recent floods

TABLE 3-2 Hazards Spatial Information Available On-line from the USEPA

Data/System	Update Frequency
Permit compliance	Monthly
Toxic Release Inventory	Monthly
Resource Conservation Recovery and Information System (RCRIS)	Monthly
Biennial Reporting System (BRS)	Every 2 years
Hazardous air pollutants (AIRS Facility Subsystem)	Monthly
CERCLIS (Superfund)	Monthly
Safe Drinking Water Information System (SDWIS)	Quarterly
Grant information and control	Biweekly
Risk management plans	Nightly
Facility information	Nightly

Source: *www.epa.gov/enviro/html/update_dates.html.*

and earthquakes, the Web site permits a user to develop customized maps for local areas (based on zip code, city, or congressional district) for selected hazards. In this way, local residents can easily see the historic patterns of hazards and risks in their community.

USEPA Envirofacts Maps on Demand (www.epa.gov/enviro/html/mod/index.html)

In many ways, the USEPA has been one of the leading innovators in interative Web-based hazards mapping. The current EnviroMapper application has three different geographic levels (national, state, and county) and covers most of USEPA's spatial databases (Table 3-2). Hazardous waste sites and toxic releases are the primary hazards available for querying. Given the changing nature of reporting, most of the data used in Envirofacts are updated frequently.

Scorecard (www.scorecard.org)

Scorecard is another interactive mapping and information dissemination vehicle for hazards information. Developed and maintained by Environmental Defense, Scorecard provides easily accessible information on environmental hazards information (pollution, toxics) to local communities. Users can create their own toxic profiles of the community by zip code or searching national maps by state and county. Hazard topics include land contamination, hazardous air pollutants, chemical releases from manufacturing facilities, animal waste from factory farms, and environmental priority setting. The information presented in Scorecard is based on USEPA data. However, Scorecard is intentionally designed to facilitate public use and to communicate the risks and hazards in order to empower communities to reduce risks in their local area.

A WORD OF CAUTION ABOUT HAZARDS MAPPING

Like other technological advances, it should be remembered that a GIS is only a tool and, as such, is only as good as the data that were entered into it. In fact, a GIS can distort true relationships and is just as prone to bias as other data-based systems. This distortion is a function of the subjectivity in the data selection and input as well as the quality of the data themselves. For example, we cannot always quantify factors (such as indirect economic damages) that might be important in hazards management and thus these would be excluded. Often, a GIS is based on available data, not what are the best data. A further distortion occurs as a consequence of moving among map scales and between different projections. This introduces a type of spatial error into a GIS.

It is impossible to represent all geographic information on a map, and so, out of necessity, some of the data are generalized or simplified. All maps (including those created using a GIS) are generalizations of the real world, and as the scale increases, more detail can be included. For example, the lines that are placed on FIRMs to delineate floodplain boundaries or flood zones are themselves generalizations. They represent the best science at the time of their construction on the nature of potential flooding in an area, but should not be construed as "absolute" truth, since we can lie just as easily with maps as we can with statistics (Monmonier 1991).

Map generalization, scale, and the initial data used to create the maps (hard copy or digital) are important to remember when interpreting maps. Although a map can convey useful information and illustrate geographic patterns, we must be aware that what we are viewing is a representation of reality that may not be equally shared by everyone. Since a map (hard copy or digital version) is an abstraction of reality, it reflects its creator's view of the real world, which is subsequently affected by the person's education, race, gender, or professional occupation. All of these factors influence our perceptions of risk and therefore how we ultimately make decisions about the management of hazards.

GIS and hazard research approaches have, at least in the past 40 years, coevolved. The early approaches to studying hazards with a GIS were often limited by the complexity of the subject matter, slow computers, and cumbersome software. The possibilities for modeling physical and social processes with a GIS are increasing. Because of such capabilities, it is increasingly unlikely that future hazard studies will be conducted without a GIS. The limitations we face today and in the future are, in large part, conceptual rather than technical, and are strongly constrained by the quality of hazard and loss data, the topic of the next chapter.

CHAPTER 4

Data, Data Everywhere, But Can We Really Use Them?

Deborah S.K. Thomas

Arriving at estimates for the frequency of hazard events or assessing loss is entirely data dependent. Unfortunately, no comprehensive database or inventory currently exists that provides an account of hazard events and losses throughout the world, much less in the United States (NRC 1999a, b). True, a plethora of data can be found, sometimes as easily as downloading it from or viewing it on the Internet. Some agencies collect data about the characteristics of specific hazards, whereas others are concerned with the effects of hazard events on people. Private companies collect data on insured losses, but these are not readily available to public agencies or researchers, nor are they necessarily geographic in nature. Natural hazards are catalogued separately from technological hazards; the same is true of climatic versus geologic events. The result is a fragmented and questionable data inventory on the nature of hazard events and the impacts that they have at the local, state, and national levels.

Using data throughout the disaster cycle to improve planning, response, relief, and recovery so that losses are minimized translates into two broad issues: (1) a seeming lack of coordination on what information is collected and (2) the need to transform data into useful and timely information that users (including those in decision-making positions) can access easily. Even though data collection efforts, databases, and data integration may not be the most fascinating subjects, a thorough appraisal of data reveals what we currently can and cannot say about hazard events and losses. This chapter explores how hazard data are collected, offers a critique of these sources, and suggests future directions for improvement.

NEED FOR SYSTEMATIC PRIMARY DATA COLLECTION

The interpretation of hazard events and loss measurements is problematic because the data are collected at a variety of scales and must be considered within the context of their use (Mileti 1999). The methods used by different agencies and organizations for collecting loss data vary widely, consider different types of costs and losses, and frequently change, making geographic comparisons over time challenging, if not impossible. Agencies have different missions, which are reflected in the type of data they collect, compile, and disseminate. Some organizations are more interested in event information for forecasting and modeling, and thus do not even collect figures on damages, economic losses, or human casualties. On the other hand, some agencies are concerned primarily with loss information but do not focus on event frequency or risk estimation. When it is included, damage information is reported in various ways as well. For example, some databases, such as those for tornadoes or lightning, report damage by category in earlier years and as actual dollar amounts beginning in the 1990s. The historical time frame for most data sets also varies substantially, limiting evaluations over time. The ability to compare geographic regions is virtually impossible because of the spatial scale at which data are collected (precise location using latitude/longitude, census tracts, counties, or state).

The need for coordinating disaster data collection and dissemination efforts has not gone unnoticed. The late 1990s saw a coordinated effort to establish a disaster information network when the Disaster Information Task Force (DITF 1997) recommended that the United States take the lead in creating the Global Disaster Information Network (GDIN). The President signed an executive order in April 2000 officially creating GDIN. Efforts are currently under way to formulate future directions for this network. Fortunately, this challenging task has many resources from which to draw. Most relevant, the Federal Geographic Data Committee is a forum in which many federal agencies have worked together since

1990 to establish the best and most efficient means for data transfer between organizations and individuals at the local, state, and national levels (FGDC 1997). The National Academy of Public Administration (NAPA 1998) endorsed the need for data development and sharing now, with more concerted efforts in the future. Of course, hazards data are only one small category of the geographic data being discussed in these forums, but the issues are no less relevant to the hazards community. Finally, consistently collecting loss data is only one component of a much broader endeavor for improved information gathering and distribution for the purpose of reducing the impacts of environmental hazards. Only time will reveal how successful GDIN will be and how the efforts already under way will contribute to its success or failure.

CENTRALIZED DATA AND INFORMATION DISSEMINATION

Many organizations have experience integrating hazard-specific information for single or multiple events, although the way they go about it can be very different. For example, a number of agencies collect and disseminate raw data (e.g., National Aeronautics and Space Administration), whereas others (e.g., National Climatic Data Center) process data and distribute summary findings. Sometimes an agency, such as the U.S. Geological Survey (USGS), fills both roles. Another approach is the idea of a data clearinghouse where data are disseminated either through downloaded files or via a series of links to the agency that actually maintains the information. This form of a distributed information system allows the clearinghouse to become a window or portal to raw data stored elsewhere. The Center for Integration of Natural Disaster Information (CINDI) at the USGS is one example. CINDI currently functions as a clearinghouse for disaster information, organizing references to data maintained by others, not only the USGS.

A larger issue is that some agencies or organizations focus on the collection and dissemination only of specific categories of hazard data, rather than all types of hazards. This is to be expected, given the missions of various agencies. For example, the National Climate Data Center (NCDC) is the repository for all types of data related to weather and climate in the United States and worldwide. Geophysical hazards, however, are not represented. Data on geophysical and geohydrological hazards are housed at the USGS. For flood events, one must go to a number of different sources to obtain flood hazard events and losses (NCDC,

USGS, Federal Emergency Management Agency [FEMA]). Technological hazards data are found predominately with the U.S. Environmental Protection Agency (USEPA) with the exception of nuclear hazards, which can be found in the U.S. Department of Energy and the U.S. Nuclear Regulatory Commission (USNRC).

Although there may be integration of broad categories of data (e.g., weather-related hazards) within one agency, a true systematic integration of multiple types of hazard data currently does not exist. Further, loss information is not always included in many of these data sets. The Center for Research on the Epidemiology of Disasters and the Office of U.S. Foreign Disaster Assistance's EM-DAT is one attempt to create an international database on disaster impacts at the country level (CRED 2000). The scale is coarser than that required for local-level assessments, but it is a start toward a spatially integrated disaster event and loss database.

Efforts are under way by some groups to pull data together on multiple hazards from various sources through geographic referencing and mapping. In fact, hazards data are increasingly available in Geographic Information Systems (GIS) formats, as noted in the previous chapter. However, most of those efforts focus on risks from single hazards through the delineation of risk zones. There is little data on multiple hazards (and thus overlapping risk zones), or on the losses attributed to specific hazard events.

The insurance industry has recognized the need for more comprehensive hazards data. At least two major efforts are under way to aggregate data on insured losses from all types of hazards. The Institute of Business and Home Safety began development of a Catastrophe Paid Loss Database in 1994 to establish loss estimates throughout the insurance industry (IBHS 1998). The Munich Insurance Group aggregates estimates on the cost of disasters throughout the world, which is summarized in a world map of natural disasters (Munich Insurance Group 1998).

WHY DATA MAY NOT BE MEANINGFUL, EVEN IF COLLECTED

Although many attempts to integrate hazards data exist, the data sources and applications do not always present a comprehensive picture of damages or loss of life. Simply put, measuring losses for hazards and disasters is a difficult proposition; and no widely accepted framework or formula for estimating losses exists (NRC 1999b). Therefore, the losses from natural and technological hazards in the United States are not known with much certainty at all.

At least four challenges are presented when trying to identify and quantify hazard losses: hazards produce multiple-order impacts, costs may be direct or indirect, quantifying intangibles is difficult, and tallying up losses against benefits is not so clear-cut (Pielke 1997). As these four problems attest, hazard loss is a much broader concept than costs or expenditures. Costs generally refer to those losses reimbursed by government or insurance, whereas expenditures involve up-front investments associated with mitigation efforts (NRC 1999b, Godschalk et al. 1999). Clearly, there are many aspects of loss that do not meet these criteria. As a result, the compilation of loss data using bits and pieces of disparate information may not fully reveal the true character of loss.

Most reported loss estimates comprise direct losses in the form of costs, rather than expenditures. **Direct loss** corresponds closely to the actual event where loss of life or the physical loss of property is apparent (Heinz Center 2000a). These losses may or may not be insured. On the other hand, **indirect losses** are often referred to as hidden because they appear later than the initial event and involve other social and economic sectors not immediately associated with the direct impact of the disaster. Examples of indirect losses include the disruption of tourism activity in a hurricane-devastated community or the loss of income to local workers because a manufacturing plant was damaged by a tornado. Another form of indirect economic loss that is rarely, if ever, included in databases is that from the informal economic sector. For example, the income loss of illegal immigrants who harvest crops is not normally calculated in the overall cost of a disaster (by indirect loss indicators), but the damage to crops is included. As a general statement, indirect losses are uninsured, a condition that illustrates the importance of looking beyond a single measure (such as direct loss) for calculating losses.

Another continually vexing question involves how to count death, injury, and damage, and when to stop counting. For example, insured losses from any single disaster event remained below $1 billion prior to 1989's Hurricane Hugo and the Loma Prieta earthquake (IBHS 1998). Since then, insured losses from Hurricane Andrew and the Northridge earthquake have exceeded $15 billion (Munich Re 2000). If indirect, direct, insured, and uninsured losses were all considered, that threshold (greater than $1 billion in loss) would have been reached sooner. Even with these factors taken into consideration, the question of how to assign value and then measure the value of losses—especially cultural or similar assets—still remains. The interpretation of those measurements is problematic as well since the data are collected at a variety of scales and for many different purposes.

Less tangible are the losses that are not so easily quantified. How do we adequately measure the loss of human life or the losses due to injuries in a disaster? What constitutes a loss for some, is an opportunity for another. For example, construction laborers see an increase in available work and outdated infrastructure is replaced to the benefit of an industry's productivity and market competitiveness. Furthermore, assessing loss that is emotional, psychological, ecological, or cultural often does not involve a simple count as one might undertake for the number of damaged structures. We need better tools and methods for estimating indirect losses.

Loss estimates also assume that people are impacted equally. In fact, many studies suggest that women, children, elderly, minority, and low-income populations are disproportionately affected by disasters (Cutter 1995, Enarson and Morrow 1998, Mileti 1999). Yet, of the loss databases that do exist, few, if any, systematically collect data on these vulnerable groups. Those who often endure the greatest burden of disasters are simply not distinguished, or become silent victims. If we are to understand all facets of loss so that we can reduce it in the future, then we must quantify the impacts on diverse subpopulations in order to better understand the differential burdens of loss.

ISSUES OF DATA SHARING

The potential for sharing data is immense and has many benefits. More and more agencies and institutions are placing data on the Internet, either in the form of raw data for download or as informational maps. Access to data is easier now, more than ever before. Clearinghouses act as an informational conduit between departments in the same agency or between agencies.

Despite the improved availability, all data have their limitations and people who use some of the national data for loss assessment should be cognizant of these caveats. All data are not of equal quality. Particularly with maps and data in digital form, people tend to lose sight of the fact that data can contain geographic or spatial errors as discussed in the previous chapter. For example, if the coordinates (longitude/latitude) for a tornado were entered incorrectly in 1965, this error now appears as a point location in the historical database, just like all other tornadoes. There is no real way to verify the location and it now appears "correct." The point location also may be in the wrong place because the instrumentation in the mid-1960s was not nearly as precise as current technologies. Misunderstandings can occur because people do not completely comprehend the nuances in the data being viewed. For instance, the designation of the 100-year floodplain may be misleading to a homeowner because it implies that by being just outside the "line" drawn on the map, there is no flood risk. Just because data are mapped does not make them any more correct.

Along with issues of accuracy and precision, organizations have other considerations where data are concerned. Data collection, maintenance, revision, and archiving are a significant financial investment. As such, data become a commodity, have a monetary value, and can be sold or purchased. The same data set can also serve very different purposes, depending on the person or agency using it. Whereas the emergency planner in a municipality may intend to use a risk map for mitigation purposes, an insurance company may be interested in raising or lowering insurance rates. Many of these issues can affect the availability of data since agencies must take them into consideration when disseminating information. There are also important legal issues and concerns (liability) about data use and interpretation that are discussed elsewhere (NAPA 1999).

AVAILABLE DATABASES FOR ESTIMATING LOSS

Any attempt at loss estimation, regardless of whether it is local or national, requires a thorough understanding of data nuances. Many of the general caveats regarding hazards data have been mentioned already. Because this book is focused on the availability of loss data associated with hazards, not with warning, prediction, or event information, some of the specific data sources that were used to describe the temporal and spatial trends in hazards losses (Chapters 5 and 6) are highlighted next. Table 4-1 provides an overview of these sources and their limitations.

Weather-Related Hazards

The NCDC's *Storm Data and Unusual Weather Phenomena* is one of the primary data sources that compiles death and damage estimates. This publication is a fairly comprehensive source for information on

TABLE 4-1 Summary of Data Sources

Hazard	Source	Dates Covered	Variables	Database Limitations
Tornadoes	Storm Prediction Center, Norman, OK, *http://www.spc.noaa.gov/archive/index.html* After 1995: National Climatic Data Center, Asheville, NC, *http://www.ncdc.noaa.gov/*	1959-present	Date, time, latitude, longitude, deaths, injuries, damage category	Limited historical data time frame; difficulty assigning damage, death, and injury to specific location
Thunderstorm wind	Storm Prediction Center, Norman, OK, *http://www.spc.noaa.gov/archive/index.html* After 1995: National Climatic Data Center, Asheville, NC, *http://www.ncdc.noaa.gov/*	1959-present	Date, time, latitude, longitude, deaths, injuries, damage category	Limited historical data time frame; difficulty assigning damage, death, and injury to specific location
Hail	Storm Prediction Center, Norman, OK, *http://www.spc.noaa.gov/archive/index.html* After 1995: National Climatic Data Center, Asheville, NC, *http://www.ncdc.noaa.gov/*	1959-present	Date, time, latitude, longitude, deaths, injuries, damage category	Limited historical data time frame; difficulty assigning damage, death, and injury to specific location
Lightning	National Climatic Data Center, Asheville, NC, *http://www.ncdc.noaa.gov/*	1959-present	Date, time, injuries, deaths, damage, location of strike, county of strike	Event only by county and state; only those events with death, damage, or injury
Storm Data (Meteorological events: wind, hail, lightning, winter hazards, tornadoes, flooding, drought, landslides, hurricanes, wildfires, thunderstorms)	National Climatic Data Center, Asheville, NC, Entered by Natural Hazards Center, Boulder, CO (1975-1994) National Climatic Data Center, Asheville, NC, (1993-present) *http://www.ncdc.noaa.gov/*	1959-present	Date, time, location, deaths, Injuries, damage category (property and crop) for multiple weather-related hazards such as flood or drought	Event-based entry impedes large-scale analysis (county or subcounty)

Hurricanes, Atlantic	National Hurricane Center, Colorado State University, *http://www.nhc.noaa.gov/pastall.html* *Monthly Weather Review*	1886-1996	Date, time, wind speed, pressure, deaths, damage	Difficulty assigning damage, death, and injury to specific location
Hurricanes, Pacific	National Hurricane Center, Colorado State University, *http://www.nhc.noaa.gov/pastall.html* *Monthly Weather Review*	1949-1996	Date, time, wind speed, pressure, deaths, damage	Limited historical data time frame; difficulty assigning damage, death, and injury to specific location
Floods	National Weather Service, *http://www.nws.noaa.gov/oh/hic/index.html*	1903-present (yearly losses) 1955-present (by state)	Year, state, death, damage	Estimates only available at the state level; no injury included
Earthquakes, Epicenter	Council of National Seismic Systems, *http://quake.geo.berkeley.edu/cnss/*	1970-present	Time, latitude, longitude, depth, magnitude	No deaths, injuries, damages
Catalog of Significant Earthquakes	National Geophysical Data Center, *http://www.ngdc.noaa.gov/seg/hazard/eqint.html*	2150 B.C.-present	Date, time, latitude, longitude, magnitude (Richter), intensity, death, damage category	Only has damage greater than $1 million; ten or more deaths; magnitude 7.5 or greater; intensity of X or greater
Earthquakes, Significant	Earthquake Research Institute, University of Tokyo, Japan, *http://www.eic.eri.u-tokyo.ac.jp/CATALOG*	3000 B.C.-1994	Date, time, latitude, longitude, magnitude (moment), intensity, death, injuries, damage category	Only earthquakes with death, injuries, or damage

continues

TABLE 4-1 Continued

Hazard	Source	Dates Covered	Variables	Database Limitations
Volcanoes	Global Volcanism Program, Smithsonian Institution, *http://nmnhgoph.si.edu/gvp/index.htm*	Approx. 8,000 B.C.-present	Volcano number, volcano name, regional location, latitude, longitude, elevation, type (morphology), status, last eruption time frame	No deaths, injuries, damages
Hazardous Materials Spill	U. S. Department of Transportation, Hazardous Materials Information System	1970-present	State, injuries, death, damage	Only by state
Hazardous Sites	Environmental Protection Agency, *http://www.epa.gov/superfund/index.htm*	1966-present	ID, name, latitude, longitude	No deaths, injuries, damages
Toxic Chemical Releases	Environmental Protection Agency, *http://www.epa.gov/opptintr/tri/*	1987-present	Facility information, amount of release, chemical released, media of release	No deaths, injuries, damages

meteorological hazards, including severe storms, tornadoes, wind, hail, ice, extreme cold, extreme heat, drought, and all other climate-related events. Each chronological entry listed by state contains information on the hazard type, death, injury, property damage, crop damage, description of the event, and an inventory of affected counties or regions within the state.

The use of *Storm Data* for hazard loss assessment has several limitations, however. First, because the data are not directly linked to any geographic database, the capability for geographic analysis is severely restricted. Hail, thunderstorm wind, and tornadoes all have latitude/longitude locations, but no other hazard events have such a precise location associated with them. Instead, a single county, multiple counties, or region of a state (e.g., northwest Kansas) is listed. With some modification, those events for which counties are listed could be mapped, but if this is not listed, only state-level mapping is possible. Also, many events transcend a political boundary, creating other difficulties in distributing damages across counties. Another limitation is the inconsistency of hazard classification. For example, snow-related events have as many as 15 different listings, ranging from blizzard/heavy snow to heavy snow to heavy blowing snow. Further complicating its use for damage assessment is the fact that, until recently, property and crop damage was entered in broad-range categories rather than actual amounts. Although *Storm Data* has many limitations when it is used for hazards assessment, it is often the only source for event and loss estimations collected in any systematic fashion at the national scale for many hazards.

Although *Storm Data* began in 1959, most of the pre-1993 information is in hard copy only. The *NCDC Storm Event Database* is a digital version of the *Storm Data* that began in 1993. As long as the digital version continues, it will be an incredibly useful source into the future and will enable users to develop more place-based studies of weather events and losses.

Hurricanes

The National Hurricane Center (NHC) rigorously collects data on hurricanes, tropical storms, and tropical depressions in order to improve warnings and reduce the loss of life. The event information contained in the databases that the NHC maintains is extensive. Data, such as central pressure, wind speed, and category, are collected for each named storm at 6-hour intervals along its path. This event data set does not, however, contain information on losses, but the NHC does release preliminary reports on each hurricane that does contain overall death and damage estimates. An annual article in the *Monthly Weather Review* describes the previous year's hurricane season. This publication summarizes the past season and then provides accounts of individual storms, including information on death, injury, and property damage for each event. This data set is particularly good for looking at yearly trends. Unfortunately, these numbers cannot always be assigned to a given state. Consequently, loss information on hurricanes by state must be culled partially from *Storm Data*, which has all of the same limitations for hurricanes as for other data sets, described previously. Still, it does provide some idea of the geographic distribution of the losses by state.

Floods

Flooding has the most diverse set of data sources available. However, these are consistently ineffectual in trying to understand the magnitude of hazard losses and the geographic distributions of events and losses. For example, the USGS maintains stream gauge readings (in near real time) that can be obtained in digital form (*http://water.usgs.gov*) but flood stage by stream is not as readily available. The USGS is not responsible for declaring flood watches and warnings. Instead, the information is passed on to the National Weather Service (NWS) and the interpretation of stream gauge data occurs there. The NWS does keep a record of associated losses with flood events, although it is not mandated to do so. Unfortunately, estimates are only maintained at the state level by year back to 1955, although national estimates go back to 1903. The U.S. Army Corps of Engineers also collects loss information, but for a much more limited time period. Further, these estimates are reported by federal fiscal year (October 1-September 30), making calendar-year (the way the vast majority of data are reported) comparisons difficult, if not impossible. As discussed previously, *Storm Data* does include death, injury, and damage information for all types of flooding, and so, information obtained from the NWS was supplemented for the estimates presented in the following chapters.

Earthquakes

With the increased installation of seismographs in all parts of the world, the quality and amount of data collected for earthquake events has improved greatly (NEIC 1999). Several data sets contain variables such as epicenter location, magnitude, intensity, and/or losses. Examples include the Catalog of Significant Earthquakes from the National Geophysical Data Center (NGDC) and the Centroid Moment Tensor Catalog from Harvard Seismology. The Council of National Seismic Systems (CNSS) is a composite catalog of earthquakes created by merging earthquake catalogs of member institutions and removing duplicate events. Thirty member institutions comprise the CNSS, including public and private universities as well as state and federal agencies. This data set includes the event's time of occurrence, location in latitude and longitude, and Richter magnitude. Although this is a good database to illustrate event distributions, no deaths, injuries, or damages are available to chart or map loss trends.

The Catalog of Significant Earthquakes is a selective database containing latitude/longitude locations for earthquakes meeting one of the following conditions: damage greater than $1 million, ten or more deaths, magnitude (Richter) 7.5 or greater or Mercalli intensity X (10) or greater. Because of the subset of events, it does not represent the true number of earthquake occurrences, nor does it portray all events that caused any type of loss. The time period is quite extensive, covering centuries. However, one important limitation in the database (similar to *Storm Data*) is the reporting of economic losses by damage category rather than dollar amount (Table 4-2).

TABLE 4-2 Earthquake Damage Categories Derived from the Significant Earthquake Catalog

Damage Category	Dollar Damage Amount
Insignificant	Little observed
Some	Some observed
Limited	< $1 million
Moderate	$1 million-5 million
Considerable	$5 million-15 million
Severe	$15 million-50 million
Extreme	> $50 million

Source: National Geophysical Data Center, *http://www.ngdc.noaa.gov/seg/hazard/eqint.html.*

Volcanoes

This database from the Smithsonian Institution's Global Volcanism Program (1999) contains only location and descriptive geologic information for volcanoes active in the past 10,000 years. No deaths, injuries, or damages are available to chart or map loss trends over time or geographically. In the United States, the deaths and damages from Mt. St. Helens were included in NGDC's Catalog of Significant Earthquakes.

Hazardous Materials Spills

This is one of the few technological hazards databases that includes information on death, injury, and damage. The U.S. Department of Transportation requires that any transportation incident involving hazardous materials be reported. These data are maintained in the Hazardous Materials Information System and include transportation type, death, injury, and damage summaries by state for each year beginning in 1971 (USDOT 2000). Unfortunately, the data are self-reported and are not verified by independent sources. Consequently, the numbers are somewhat questionable, but they at least provide some basis for loss estimation comparisons over time and geographically by state.

Hazardous Waste Superfund Sites

Like much of the data on chronic technological hazards, this database also does not contain information about deaths, injuries, or damages (USEPA 2000f). The data only contain locational (latitude/longitude) information. They can be used to evaluate those who are potentially impacted by these sites, but the specific threat posed by most hazardous sites is not well documented or understood at the national level. We do know, however, of the relative risk associated with these sites based on their listing on the National Priority List. Data on the costs of remediation are available, but the temporal and spatial coverage is often limited.

Toxic Chemical Releases

The Emergency Planning and Community Right-To-Know Act (EPCRA) of 1986 (also known as Title III of the Superfund Amendments and Reauthorization Act) provides for the collection and public release of information about the presence and release of hazardous chemicals. Through EPCRA, Congress mandated that a Toxic Release Inventory (TRI) be made public to help citizens and community leaders be better informed about hazardous materials in their communities. Facilities are required to report to the USEPA and state governments on releases into the air, water, and land; they must also report the transfer of wastes for treatment or disposal at a separate facility. Note that the TRI reports reflect releases of chemicals, not exposures to people. These estimates alone cannot be used to determine potential adverse effects on human health and the environment. No information about deaths, injuries, or damages are reported (Table 4-1).

Some controversy exists over the reporting accuracy of industrial emissions generally (AWMA 1997) and TRI specifically (Lynn and Kartez 1994). The releases are self-reported annual estimates, and facilities may either over- or underestimate their releases, depending upon their estimation methodology. Some facilities fail to report at all or report only some of their covered chemicals. Additionally, the amounts reported could be the result of a single release or may have been released evenly throughout the year. Still, this particular database is commonly used to assess the differential exposures by various subpopulations, especially in environmental equity studies.

Nuclear Incidents

In the absence of a historical record of nuclear incidents comparable to other threats, such as hurricanes or earthquakes, another measure is needed. Equipment failure and human error both contribute to increasing the likelihood of an incident, and so, a review of safety records for nuclear facilities provides a picture of the relative level of risk posed by one facility over another. The *Systematic Assessment of Licensee Performance* (SALP) by the USNRC is one measure of plant safety that may be used to assess the relative risk from a particular facility (USNRC 1996).

Conducted roughly every 18 to 24 months, the SALP reports grade several operational categories at each commercial nuclear facility (Table 4-3). The USNRC has recently discontinued the SALP review as part of its development of a new reactor oversight and assessment program. The new program began on a pilot basis at nine plants in June 1999 and was adopted for all facilities early in 2000 (USNRC 2000).

TABLE 4-3 Commercial Nuclear Power Facilities Systematic Assessment of Licensee Performance

Areas Assessed	Data Years
Operations	1988-1998
Maintenance	1988-1998
Engineering	1988-1998
Plant support[a]	1988-1998
Radiological controls	1988-1993
Emergency preparedness	1988-1993
Security	1988-1993

[a]Plant support was subdivided into three separate categories, as noted, from 1988-1993.
Source: USNRC (1996).

CONCLUSION

Recently, many have recognized the distinct need to institutionalize the collection of compatible data on hazard events and losses. The extent to which losses may be mitigated effectively depends upon an understanding of current and historic trends. Losses stemming from natural and technological sources in the United States, however, are not known with certainty. The plain fact is that, unless loss assessments are based on reliable, comprehensive data collected in a systematic fashion, their value is limited.

We still do not know the true extent of hazard losses in this country. If we simply look at natural hazard events, we can speculate on the economic value of direct losses for a given year, but this is an educated guess, not based on strong empirical proof. We are less able to determine what natural hazards cost at the county or community level and their local economic impact. Only through a systematic effort of loss estimation procedures, collection of comparable data across hazards, georeferencing of all data, and the archiving of the resulting databases will we be able to assess our collective progress toward hazard loss reductions. The time for a multihazard national inventory of hazard events and losses is long passed.

CHAPTER 5

Trends in Disaster Losses

Jerry T. Mitchell and Deborah S. K. Thomas

Fatalities and economic losses from disasters are rising. Although a worldwide phenomena, these losses are not uniformly distributed geographically or through time. The juxtaposition of hazard events and vulnerable populations varies spatially and, consequently, so do losses. A direct relationship exists between the level of development and the type of losses that predominately occur. In developing nations, the death toll from disasters is much higher than in more developed countries (IFRCRCS 1998, CRED 2000). In the United States and other parts of the developed world, the escalation in losses is most profound in economic terms (van der Wink et al. 1998, Munich Re 2000). What is driving this upward course of losses? If this nation is indeed becoming more disastrous, is a rise in physically damaging events responsible? Are the trends a mere artifact of better reporting of loss data? Is it a symptom of ill-regarded choices made through our social, political, and economic systems? Or, is it some combination of all of these?

As we mentioned in the previous chapter, the compilation of loss data is fraught with difficulties. The most significant is that the data we have only report a portion of the true losses (damages from the events as well as costs associated with relief and recovery operations) from natural hazards. The hidden costs (indirect costs, beneficial uses of hazardous areas that might ameliorate losses, and intangible losses) are rarely included in official tallies of what disasters cost this nation. Thus, the empirical support for understanding economic losses due to disasters, at this point in time, is based solely on directly measurable damages, and sometimes relief and recovery costs associated with specific hazard events. Oftentimes, if we cannot put a dollar figure on the impact, then it is not included in damage estimates or response costs associated with natural disasters. Therefore, any database on disaster losses, out of necessity, **underestimates** the total economic and social impact of natural hazards to communities, individual states, and the nation.

This chapter presents an empirical analysis of losses from natural disasters (and a few technological ones where possible). Hazard events, economic losses, and human casualties were compiled for the period 1975-1998 to assess changes in trends in losses over time (this chapter) and then geographically (Chapter 6). Many of the caveats enumerated above and in the previous chapter also apply to our data as well. As we have cautioned in the previous chapter, the dollar losses we report may not be consistent with losses reported from other sources, especially for some of the large, singular events. Nevertheless, the database used in this analysis is the most comprehensive and complete assessment of natural hazard losses in the nation at this time. The next section describes the natural hazard events loss database. The rest of the chapter describes the trends in events and losses from 1975 through 1998 based on these data.

BUILDING THE HAZARD EVENTS AND LOSSES DATABASE

In conjunction with the Natural Hazards Research and Applications Information Center at the University of Colorado, we compiled a subset of the National Weather Service's (NWS) *Storm Data and Unusual Weather Phenomena* as part of the second assessment of natural hazards in the United States (Mileti 1999). Only those single events with at least $50,000 in damages for 1975-1994 were included. Although not ideal for the analyses of events, this initial data set at least provides additional loss information on a wide variety of hazards. The sections on floods, thunderstorm winds, tornadoes, hail, drought, winter hazards, wildfire, and lightning draw on this database. The remaining hazards and updates to 1998 are based on the data sources described in Chapter 4.

In some data sets, damage was reported as an actual dollar amount, whereas in other databases damage was given as a category. All data that were reported as damage categories in any of the databases were conservatively assigned an actual dollar value based on the lowest dollar figure in the range provided. For example, in the National Climatic Data Center's pre-1990 *Storm Data*, a damage category of 5 represented damage ranging from $50,001 to $500,000. For our purposes, a figure of $50,001 was assigned to that event. Consequently, all damage summaries presented in the following two chapters are extremely conservative, representing the lowest damage scenario rather than the highest. In all likelihood, overall damages are much more extreme than presented here. All damages were also adjusted to 1999 U.S. dollars to account for inflation and to allow for comparisons over time (American Institute for Economic Research 1999).

No attempt was made to independently verify economic loss data, death or injury statistics. We assumed that the databases used were fundamentally correct. We know, however, that in many instances they represent underreporting in all categories and should be considered an extremely conservative estimate of the nature of losses resulting from environmental hazards in this country.

HISTORICAL LOSSES FROM HAZARDS

According to our conservative estimates, environmental hazards caused over $300 billion in property and crop damage and nearly 9,000 deaths during the 1975-1998 study period. The historic impact of individual hazards is shown in Table 5-1. Floods—coastal, riverine, and flash—were the most damaging and deadly threat with a total of $106 billion in damages and over 2,400 deaths during our study period. Although not producing nearly as much damage, lightning proved to be the second leading cause of disaster-related deaths during this time period. Tornadoes and severe winter storms also caused significant loss of life, with over 1,300 and 1,000 fatalities, respectively—or roughly 100 people per year. Regarding damages, hurricanes had the second highest economic impact at over $75 billion. Perhaps somewhat surprising, tornado damages surpassed earthquake damages ($36 billion vs. $31 billion) during the past 24 years. Severe winter storms also caused a notable amount of damage, around $20 billion.

Loss of life averaged 375 people per year for this study period. The deadliest year was 1980, attributable in part to nearly 400 deaths result-

TABLE 5-1 Summary of Hazard Impacts, 1975-1998 (Damages Adjusted to U.S. $1999)

Hazard	Events	Deaths	Injuries	Damage (millions of dollars)	Average Annual Losses (millions of dollars)
Drought	na[a]	0	0	14,693.7	612.2
Earthquakes[b]	784,439	149	na	31,454.4	1,310.6
Extreme cold	na	228	406	2,847.7	118.7
Extreme heat	na	566	1,328	1,048.1	43.7
Floods	na	2,495	na	105,868.0	4,411.2
Hail	103,243	15	569	4,863.7	202.7
Hazardous materials	259,384	580	12,897	775.5	32.3
Hurricanes[c]	82	394	4,026	75,717.7	3,154.9
Lightning	na	1,667	7,566	604.1	25.2
Tornadoes	22,409	1,344	29,437	36,627.3	1,526.1
Volcano[d]	na	32	na	2,221.0	92.5
Wildfires	na	10	278	1,532.6	63.9
Wind	126,667	470	5,628	4,002.7	166.8
Winter hazards	na	1,049	11,364	19,931.3	830.5
TOTAL	na	8,999	73,499	302,187.8	12,591.3

[a]na = not available

[b]Earthquake epicenters falling within state boundaries; there were 45 considered "significant" (see text for definition).

[c]Includes any storm tract collected by the National Hurricane Center that made landfall in the United States 1975-1998; injuries were derived from *Storm Data*.

[d]This only includes eruptions of Mt. St. Helens, Washington, Kilauea, Hawaii, and Redoubt, Alaska.

ing from extreme heat during the summer months (Figure 5-1a). Another peak year, 1983, saw multiple hazards, such as floods and extreme heat and cold, contributing to relatively high death tolls. In terms of damages, losses have climbed over time as a direct result of large-scale events, with new loss records set in 1989 and 1992 and with the 1989 record exceeded in 1993 and 1994 as well (Figure 5-1b). During the 1980s, the costliest disaster year in 1989 closed out the decade when both Hurricane Hugo and the Loma Prieta earthquake struck. Three years later, Hurricane Andrew raced over Florida, helping 1992 to become the costliest disaster year ever. The Midwest floods (a 500-year event) and a severe blizzard followed in 1993. The Northridge earthquake struck in 1994, ranking among the most costly single disaster events in U.S. history, according to our data (Table 5-2). As can be seen, however, the

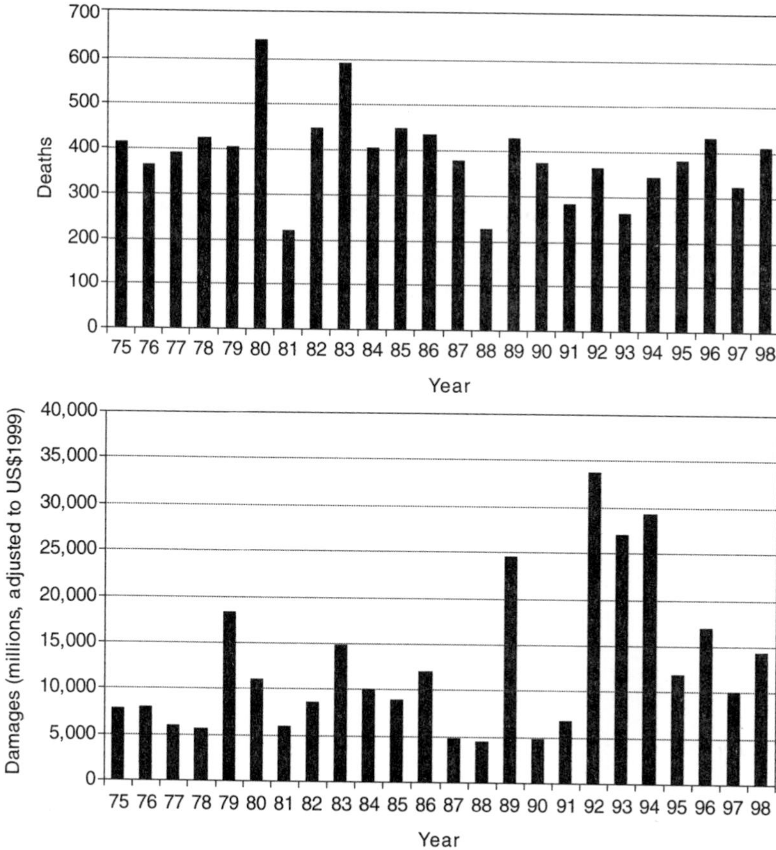

FIGURE 5-1 Trends in losses from all hazards, 1975-1998: (a) deaths and (b) damages (in 1999 dollars).

ranking of "most disastrous" or "costliest" depends on the source of the data and the definition of cost—issues discussed in the previous chapter.

Clearly, individual disasters can have a huge impact on overall losses. In fact, six events caused approximately one-quarter of all damages during this period. Although not falling within our study period, Hurricane Floyd, 1999's major entry, continues this trend. Passing relatively close to the entire U.S. East Coast, the hurricane brought high rainfall totals that produced an inland flood disaster, with damage estimates around $6 billion and 77 deaths in the United States (NCDC 2000). Hurricane Floyd was the deadliest U.S. hurricane since Agnes in 1972. Although these types of events grab our attention, we must still recognize that

TABLE 5-2 Rankings of Costliest Disasters by Different Sources

Billion Dollar Weather Disasters (1980-1999)[a]			Presidential Disaster Declarations (1988-1995)[b]		
Date	Loss (billions of dollars)	Event	Date	Loss (billions of dollars)	Event
Summer 1988	40.0	Drought/ heat wave	Jan. 1994	3.3	Northridge earthquake
Aug. 1992	27.0	Hurricane Andrew	Aug. 1992	1.64	Hurricane Andrew
Summer 1993	21.0	Midwest floods	Sept. 1989	1.26	Hurricane Hugo
Summer 1980	20.0	Drought/ heat wave	Summer 1993	0.87	Midwest floods
Sept. 1989	9.0	Hurricane Hugo	Oct. 1989	0.76	Loma Prieta earthquake
Summer 1998	6.0	Drought/ heat wave	Sept. 1992	0.24	Hurricane Iniki
Sept. 1999	6.0	Hurricane Floyd	July 1994	0.24	Southern severe storms
May 1995	5.0-6.0	Southern severe weather/flood	Feb. 1993	0.20	California floods
Fall 1995-Summer 1996	5.0	Southern Great Plains drought	Jan. 1995	0.18	California severe storms
Sept. 1996	5.0	Hurricane Fran	Oct. 1994	0.15	Texas floods
Sept. 1998	5.0	Hurricane Georges			

[a]NCDC (2000), *http://www.ncdc.noaa.gov/ol/reports/billionz.html.*
[b]Godschalk et al. (1999).
[c]FEMA (2000), *http://www.fema.gov/library/df_2.html.*
[d]Insurance Services Office, Inc. 2000. *http://www.iso.com/hurricane_experience/index.html.*

FEMA Relief Costs (1989-1999)[c]			Insurance Services Office Hurricane Losses (Adjusted $)[d]		
Date	Loss (billions of dollars)	Event	Date	Loss (billions of dollars)	Event
Jan. 1994	6.9	Northridge earthquake	1992	22.9	Hurricane Andrew
Sept. 1998	2.5	Hurricane Georges	1965	7.4	Hurricane Betsy
Aug. 1992	1.8	Hurricane Andrew	1989	7.4	Hurricane Hugo
Sept. 1989	1.3	Hurricane Hugo	1970	4.2	Hurricane Cecilia
Summer 1993	1.2	Midwest floods	1998	3.1	Hurricane Georges
Oct. 1989	0.87	Loma Prieta earthquake	1954	3.0	Hurricane Hazel
April-May 1997	0.73	Red River Valley floods	1979	2.8	Hurricane Frederick
Sept. 1999	0.73	Hurricane Floyd	1954	2.7	Hurricane Carol
Sept. 1996	0.62	Hurricane Fran	1960	2.7	Hurricane Donna
June-July 1994	0.54	Tropical Storm Alberto	1995	2.6	Hurricane Opal

approximately 75 percent of all losses were not caused by a singular "big event." Smaller events are important as well, because they often occur more frequently, and, along with the large-impact disasters, cumulatively define the various hazardscapes of the United States.

TRENDS IN SPECIFIC HAZARDS

Floods

Floodplains have long been attractive places for human occupancy. Some provide exceptionally rich soils for agriculture and most have relatively flat slopes suitable for easy building construction. Perhaps more important for the development of the United States was the need for cities to locate along the ocean, lakefronts, and rivers to facilitate trade and commerce. Waterborne transport was the least costly, and the benefits of hydroelectric power generation were well understood. Several early American cities not only had accessible harbors, but were also located along the Fall Line, the farthest point a river is navigable and a prime site for hydropower. Examples include Philadelphia, Pennsylvania, Baltimore, Maryland, and Raleigh, North Carolina. Prior to the introduction of railroads (a stimulus for cities such as Atlanta, Georgia), most inland settlements likewise were established near water bodies of all types.

The end result of the human-use system dependence upon water and the historic evolution of American cities is that, today, very few urban settlements are completely immune from floods. Since people have placed themselves in harm's way, and flooding to some degree is a fairly frequent occurrence, there is a high likelihood for loss from this natural hazard. Although flooding may not receive as much coverage in the media as the more explosive and dramatic events like hurricanes or earthquakes, floods are the number-one hazard in terms of property losses and human fatalities in the United States.

The total number of deaths from floods over our study period was 2,495. Unfortunately, the NWS data used for flood hazards does not include injury statistics. Figure 5-2a displays the trends in flood fatalities. The number of deaths averaged slightly more than 100 per year nationally during the 1975-1998 time frame. Two fatality troughs are evident in 1980 and 1988 (the only year with fewer than 50 deaths). Both years coincide with strong drought years, but one must be careful in trying to correlate the two hazards, especially over a large area such as

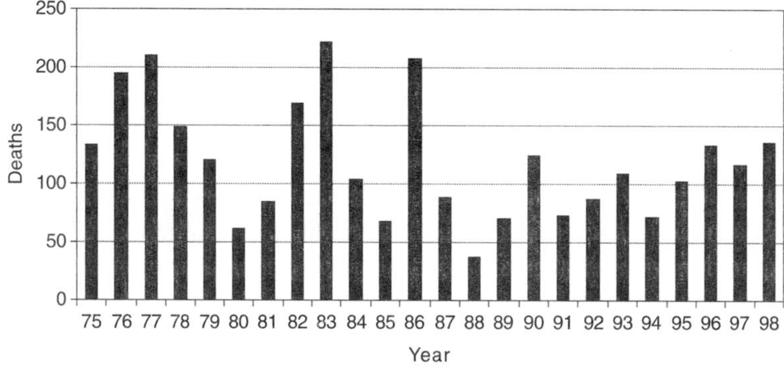

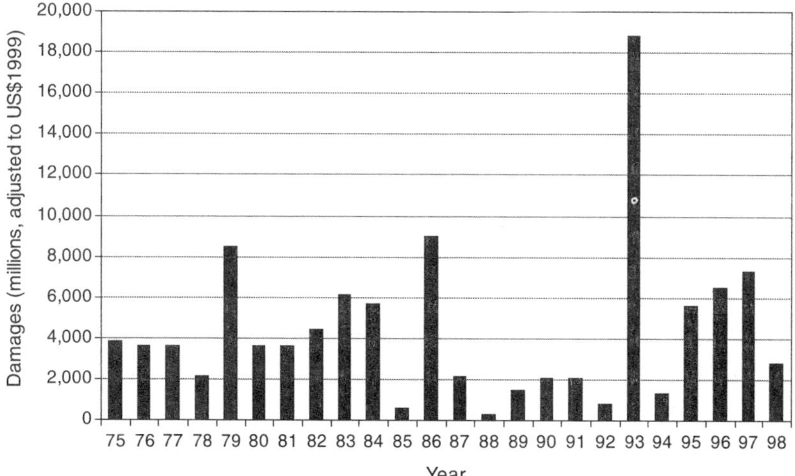

FIGURE 5-2 Trends in losses from flooding, 1975-1998: (a) deaths and (b) damages (in 1999 dollars).

the United States. For example, 1976 was a disastrous year nationally for both flood fatalities and drought losses.

Four years—1976, 1977, 1983, and 1986—closely met or exceeded 200 annual fatalities. Colorado's Big Thompson flood recorded at least 139 fatalities in 1976, comprising the majority of that year's total (Gruntfest 1996). A similar event in Rapid City, South Dakota, claimed over 230 lives just 4 years earlier. The ability of so few to survive these flash floods is indicative of the event's rapid rate of onset.

The recurrence of flooding in Johnstown, Pennsylvania, explains 1977's loss of life. Seventy-eight more deaths were added to that community's tragic flood history total, which began with the failure of the

Conemaugh River's South Fork Dam in 1889, when more than 2,200 lives were lost (McCullough 1968). A smaller flood event in the same area claimed another 25 lives in 1936 (Foote 1997). Many of 1977's remaining fatalities were a result of dam failures. Forty-three more lives were lost in Pennsylvania when the Laurel Run Dam failed and 39 people died when the Kelly Barnes Dam in Georgia failed later that year (FEMA 2000c). Multiple smaller-scale flood events cumulatively raised the death tolls for 1983 and 1986. Floods in Texas and California took the most lives in 1983 (45 and 28, respectively). West Virginia added another 38 people to the tally in 1986. Overall, our review of flood fatality data concurs with other research that found flood-related deaths were highly variable from year to year, showing neither an increasing nor a decreasing trend (L. R. Johnson Associates 1992).

Flooding is **the** major source of monetary loss in the United States from natural hazards. Losses averaged slightly more than $4.4 billion dollars annually between 1975 and 1998. This estimate is nearly double the figure reported by FEMA (1997a), which attributed an annual average of $2.15 billion for flood events during the 1951-1985 period. The three costliest years were 1979, 1986, and, of course, 1993 (Figure 5-2b).

The 1993 Midwest floods in the Upper Mississippi River basin rank as one of the most damaging natural disasters in United States history (Changnon 1996) (Figure 5-3), but certainly not the first catastrophic flood to affect America's great watercourse (Barry 1997). Although not unusual compared to annual death totals (only 38-47 flood-related deaths), these floods caused damages between $12 billion and $16 billion over a large portion of the central United States (IFMRC 1994). The relatively low loss of life for an event of this size may be attributed to its duration (from June to August), its rate of onset, and the ability to track floodwater movement successfully. Indeed, in the absence of these factors, the methodic evacuation of over 50,000 people may not have been possible (NOAA 1994). The lessons learned from this experience are enduring—future flooding is inevitable and thus we need to restrict development in floodplains and develop sustainable approaches to flood hazard mitigation (Changnon 1996).

Tornadoes

Whether single funnel clouds or a part of a larger outbreak, tornadoes are the most violent and damaging of the multiple weather hazards generated by thunderstorms. During the entire 24-year study period,

FIGURE 5-3 View of 1993 Midwest floods from bridges crossing the Mississippi River at Quincy, Illinois. Source: FEMA; photograph by Andrea Booher.

approximately 22,000 tornado events (Figure 5-4a) caused 1,300 deaths, 29,000 injuries, and over $36 billion in damage. From 1975 to 1998, there was an average of 58 deaths, 1,300 injuries, and $1.5 billion in damage per year. Although other sources provide higher death statistics—234 deaths per year between 1916 and 1950 (Bryant 1991) and 96 per year between 1953 and 1989 (Golden and Snow 1991), our frequency and mortality data are consistent with records of the more recent period (Fujita 1987, Grazulis 1993).

Casualties (deaths and injuries) reached their highest peaks in 1979, 1984, 1985, and again in 1998 (Figure 5-4b). A twister—at times as large as a mile wide—caused more than 40 deaths and 1,700 injuries in 1979 as it made its way through Wichita Falls, Texas (NWS 2000). Several causalities occurred when the storm hit a shopping mall. "Super" outbreak events, smaller than the major 1974 Midwest event of the same name that claimed 315 lives, were responsible for the fatalities and injuries in 1984 and 1985. Collectively, both North Carolina and South Carolina were hit by 22 tornadoes, resulting in 57 deaths in late March 1984. In May 1985, 41 tornadoes struck Pennsylvania and Ohio, adding 75 more fatalities to the yearly total (FEMA 1997a).

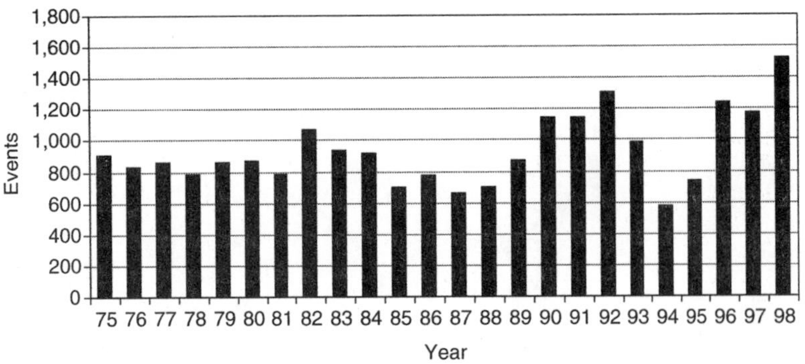

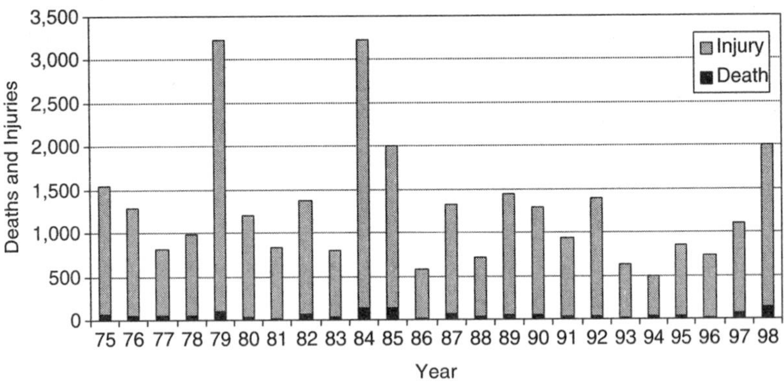

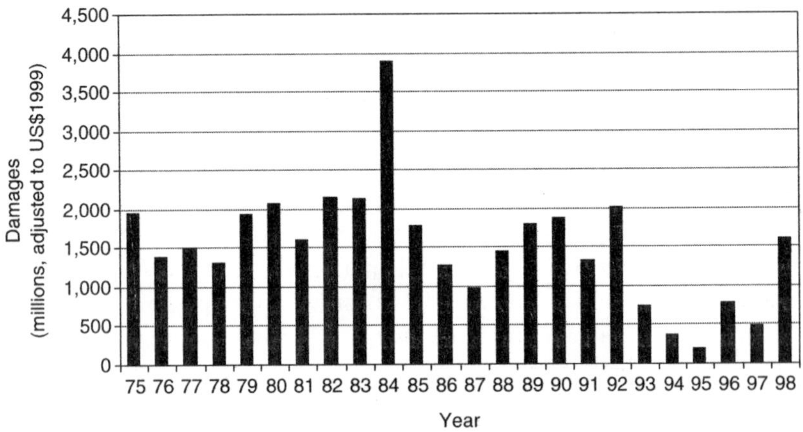

FIGURE 5-4 Trends in tornado events and losses, 1975-1998: (a) events, (b) deaths and injuries, and (c) damages (in 1999 dollars).

The spring of 1998 also was devastating, with six southeastern states suffering greatly. Relatively speaking, North Carolina, Kentucky, and Georgia fared better than most in the region, with 18 fatalities. Alabama had almost twice as many (34) fatalities and more than 700 injuries when tornadoes struck areas near Birmingham and Tuscaloosa. Another 42 persons died and nearly 300 more were injured when seven tornadoes plowed through the Kissimmee, Florida, area in late February. A dramatic event that crossed downtown Nashville, Tennessee, left at least 8 dead and 155 injured. One of the more tragic outcomes of that violent spring was the decimation of Spencer, South Dakota (Figure 5-5). A late May tornado destroyed 187 homes and all of the town's businesses; out of a small-town population of only 300 people, 6 were killed and over 150 were injured. This event mirrored the complete loss of the small town Jarrell, Texas, the previous May. Of the 400 residents 28 perished and nearly all homes were destroyed.

Damage losses closely mirror the tornadic outbreaks already discussed. Recorded damages peaked in 1984 at nearly $4 billion, but appeared to decline in the 1990s, prior to the 1998 and later storms (Figure 5-4c). The disastrous end of the decade (major events in Clarksville, Ten-

FIGURE 5-5 Damage to farmhouse and surroundings from 1998 Spencer, South Dakota, tornado. Source: FEMA; photograph by Andrea Booher and Brian Hvinden, OPA.

nessee, and Oklahoma City and Moore, Oklahoma) suggests that any perceived downward loss trend is not here to stay. Throughout the study period, the overall frequency of tornadoes remained fairly constant, although there was a slight upward trend during the 1990s.

One likely explanation is that prior to Doppler detection technology, tornadoes in uninhabited areas went largely unnoticed. Another explanation centers on localized risk factors such as time of day (television and radio warning systems are less effective at night when residents are asleep), type of dwelling (mobile homes) or shelter, and age of residents (Schmidlin and King 1995, 1997; Lillibridge 1997).

Hail

Most hail develops from thunderstorms in which strong vertical motions drive water droplets cyclically through the clouds. Eventually, the frozen droplets enlarge to the point that they fall from the clouds to reach the ground. The damage associated with hail depends largely upon the number and size of the hailstones and the speed at which they are driven. Few deaths have been attributed to hail, but significant damage has occurred to property (primarily automobiles) (Figure 5-6) and crops (especially long-stemmed ones such as corn). Within the United States, damaging hail events occur most often between the months of April and October.

The frequency of recorded hail events has steadily increased over the entire study period, with peaks in 1992, 1996, and 1998 (Figure 5-7a). It is unclear whether there are actually more hail events occurring, or if these figures represent artifacts of better monitoring and reporting systems. For example, the specific point location data for hail (e.g., longitude/latitude) only go back to 1983 for death and injury statistics, and to 1987 for estimated dollar damages. The earlier time periods (1975-1982) report events as county totals. Thus, we have a combination of specific and general data that were compiled into a statewide total by year.

Compared to other natural hazards, the threat to life from hail events is fairly benign. Hail accounted for only 15 fatalities over the 15-year period (Figure 5-7b). The relative number of injuries was also low—just 569 in over 103,000 events. Injuries from hail have rarely exceeded more than 30 persons per year, with the only exceptions coming in three peak years—1979, 1992, and 1996. Monetary losses were much more significant, however, averaging $203 million per year—or a total of $4.9 billion between 1975 and 1998 (Figure 5-7c). Another estimate places the

FIGURE 5-6 Severe wind driven hail damage to a house in St. Nazianz, Wisconsin, from a supercell thunderstorm on May 12, 2000. Source: *http://www.crh.noaa.gov/grb/may1200svr.html.*

annual loss figure at nearly $1 billion (NWS 1994). Unlike other environmental threats, there are rarely large-scale hail disasters. Losses from hail tend to be cumulative, adding up from thousands of events as each year goes by. Large-scale, damaging events have been noted, however, with the most recent occurring in Denver, Colorado (1994), and eastern Oklahoma and Texas (1995) (FEMA 1997a).

Wind

Damaging wind events other than tornadoes occur in all areas of the United States. More localized wind events also transpire in areas with unique climatic patterns influenced by mountainous terrain or the moderating influence of large bodies of water. Coastal areas exposed to tropical cyclones are among the most vulnerable to wind damage. Damages, injuries, and the loss of life typically result from building collapse or wind-driven debris. Wind events are defined in the event loss databases as thunderstorm winds at or near 50 knots, and thus do not represent damages from all known wind events.

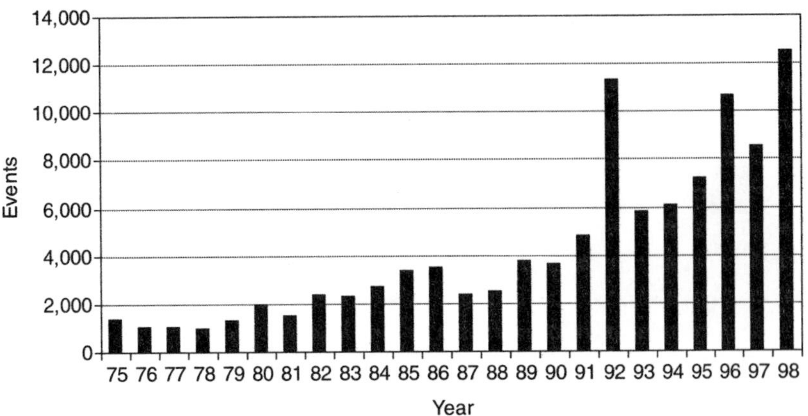

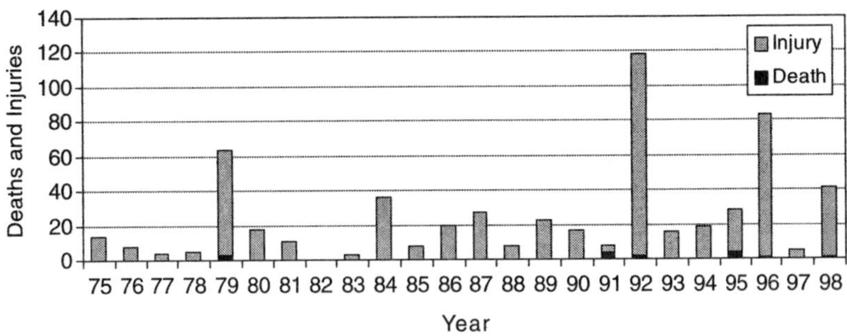

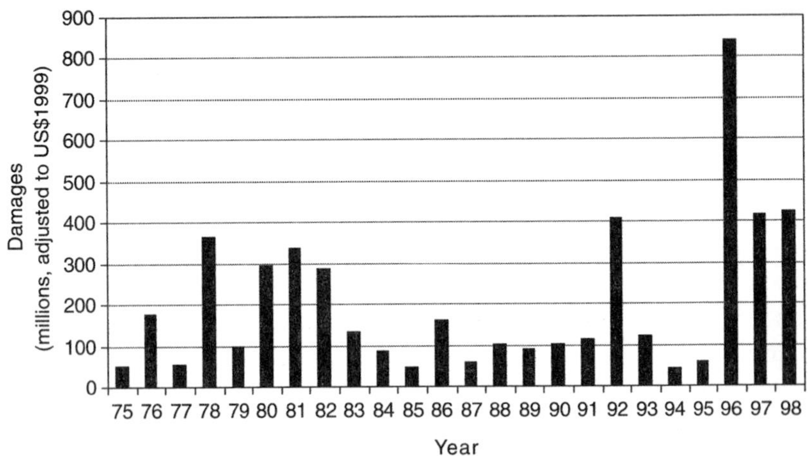

FIGURE 5-7 Trends in hail events and losses, 1975-1998: (a) events, (b) deaths and injuries, and (c) damages (in 1999 dollars).

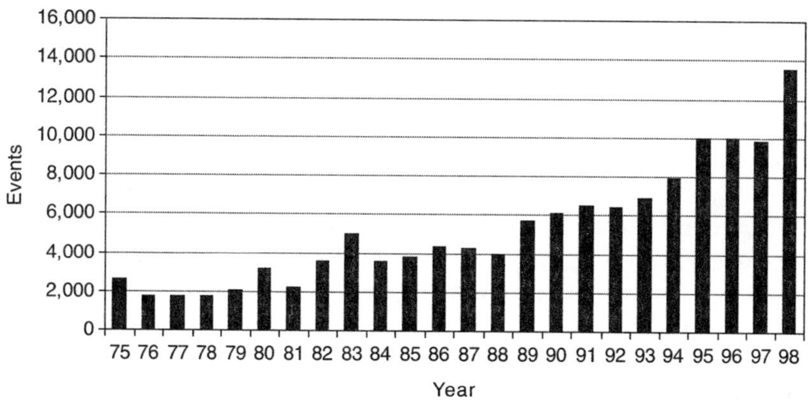

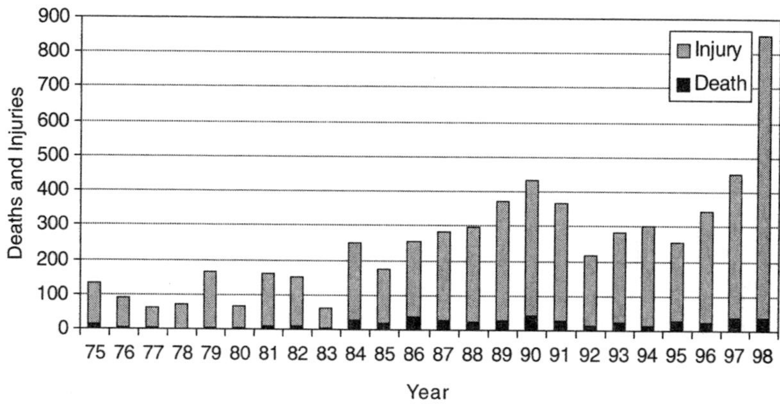

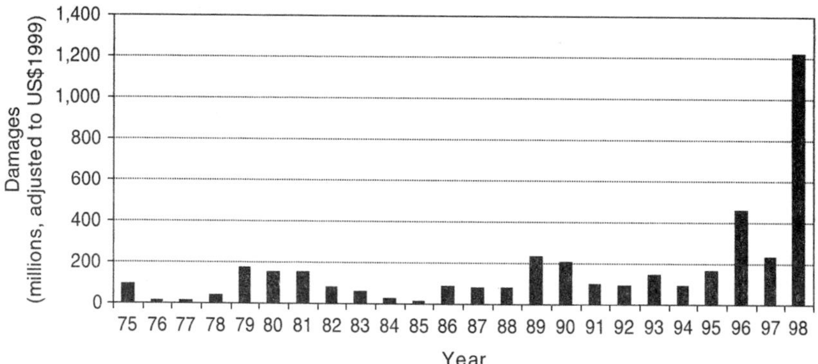

FIGURE 5-8 Trends in thunderstorm wind, 1975-1998: (a) events, (b) deaths and injuries, and (c) damages (in 1999 dollars).

The total number of thunderstorm wind events gradually increased between 1975 and 1998. The rise in events is partially due to the increased recognition and detection of this hazard and the oscillations in El Niño/La Niña cycles that contribute to severe weather in the United States. Both the greatest number of wind events (13,517; Figure 5-8a) and largest number of casualties (38 deaths and 815 injuries; Figure 5-8b) occurred in 1998, an especially strong El Niño year. The greatest amount of damage, approximately $1.2 billion, also occurred during 1998 (Figure 5-8c), a year that saw severe El Niño storms in Southern California, strong warm chinook winds in Boulder, Colorado, numerous blizzards and ice storms in the East, and severe weather in the Southeast. During the 24-year period, 20 fatalities and 235 injuries occurred annually, with yearly losses averaging approximately $167 million. With the lack of uniform building codes, the construction of housing with substandard materials, and burgeoning population growth in vulnerable areas, wind damages and casualties will continue to occur and increase in the years ahead.

Lightning

Lightning, another hazard associated with thunderstorms, is a visible flash in the atmosphere, an electrical charge that typically lasts less than a second, which can generate up to 100 million volts of electricity. Direct lightning strikes suffered by people are often fatal. The intense heat from lightning can also ignite fires that damage structures, forests, and cropland.

In contrast to other hazards, lightning fatalities tend to be individual occurrences. Exceptions, such as the loss of 38 people in a lightning-struck plane in 1963 near Elkton, Maryland, are rare. The average number of 69 fatalities and 315 injuries per year (Figure 5-9a) varies slightly from an estimate of approximately 89 fatalities and 300 injuries per year from 1963 to 1993 (NWS 1994). The number of lightning strikes that caused injury or death remained fairly consistent during the 24-year period, with peaks in 1994 and 1995. The majority of deaths and injuries occur when people become vulnerable in unsheltered areas such as parks, playgrounds, or golf courses.

It is estimated that there are more than 300,000 lightning claims filed per year in the United States with more than $33 million in estimated damages (Holle et al. 1996). Based on our data, total lightning

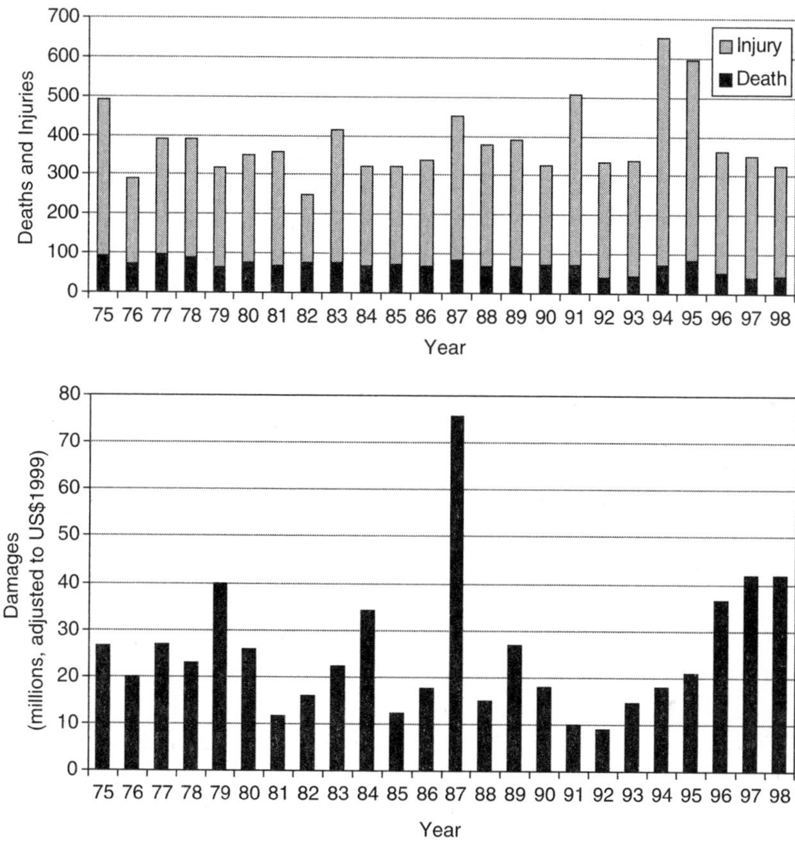

FIGURE 5-9 Trends in losses from lightning, 1975-1998: (a) deaths and injuries, and (b) damages (in 1999 dollars).

damages between 1975 and 1998 were approximately $604 million (Figure 5-9b). Although the damage impacts of lightning are not as great as some of the other hazards, the mortality statistics from lightning strikes (1,667 fatalities), makes this hazard the second most deadly after flooding. The peak in damages in 1987 is due to lightning-caused fires.

Wildfires

Wildfires have multiple causes: lightning strikes, arson, human carelessness with lit objects, and intentional burns gone awry. An example of the first cause of wildfires is the catastrophic western forest fires during the summer of 2000 (Kirn 2000). An example of the last causal mechanism for wildfires is the prescribed burn that went out of control and torched much of Los Alamos, New Mexico, in the spring of 2000. Wildfires have long been part of our cultural history and landscape (Pyne 1997), but long-term data for wildfires are much less complete than for other hazards. Some data are available for major events, but a comprehensive compilation does not exist at this time, although the National Interagency Fire Center does provide limited historical statistics on frequency of wildland fires and acres burned (NIFC 2000). For example, from 1988 to1999, there were, on average, 62,000 wildland fires that burned a total of 2.9 million acres per year (NIFC 2000). Three years stand out as high-burn years—1990, 1996, and 1999—with more than 5 million acres burned (Figure 5-10a).

In U.S. history, the most damaging wildfire event in terms of lives lost was the 1871 Peshtigo Fire in Wisconsin, in which nearly 1,200 people perished (Nash 1976). Contemporary events have been less fatal, but damages continue to increase. From 1975 to 1998, wildfire losses totaled $1.5 billion (Figure 5-10b). Other estimates—even for single events—are much higher. Substantial losses, for example, occurred as a result of the 1991 Oakland, California, fire. Damages have been calculated at more than $1.5 billion; 25 people were killed and an additional 150 were injured (COES 1994). Another $1 billion in losses (and three more fatalities) were added to the state's burden with the Southern California fires in late 1994 (FEMA 1997a).

The fact that wildfires can occur in virtually every state was demonstrated by the Florida wildfires of 1998. Wildfires were reported in more than 30 Florida counties, with Duval and Flagler near Jacksonville being among the most threatened (Figure 5-11). Portions of Interstate 95 (the main north/south route on the East Coast) had to be closed because of poor visibility from the smoke. Nearly 2,300 fires burned 500,000 acres, damaged or destroyed over 300 homes, and ruined timber valued at over $300 million (GWRMRC 1998). Major urban/wildland interface events such as those in Florida, and the rapidly developing West, appear to be increasing as people choose to live in areas with high wildfire risk.

Drought

There are several types of drought. Meteorological drought, an extended deficiency from normal precipitation levels, is the type usually considered by the public, but socioeconomic droughts (greater demand for water than supply) occur as well. Drought is intensified by high temperatures but it does not occur only in hot weather, and a heat spell does not signify a drought. Because it is a chronic and slow-onset hazard, drought is perhaps the most underrated hazard in terms of the strains it puts on the people and the economy of the United States (Wilhite 1993, 1996, 1997). One comparison of annual monetary losses from drought even ranks this threat ahead of floods and hurricanes. In a comparison of

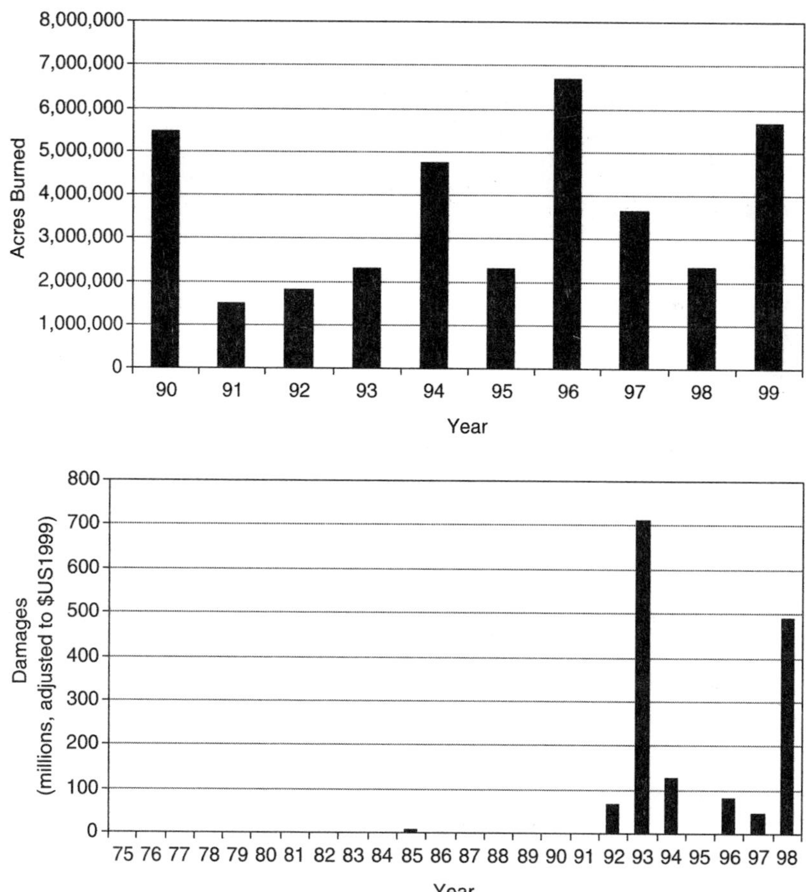

FIGURE 5-10 Trends in losses from wildfires, 1975-1998: (a) acres burned, and (b) damages (in 1999 dollars).

FIGURE 5-11 A damaged home in Flagler County, one of the hardest-hit regions during the 1998 Florida wildfires. Source: FEMA; photograph by Liz Roll, OPA.

drought, flood, and hurricane losses made by the National Drought Mitigation Center (Knutson 1997), annual losses were summed as follows: drought, \$6 billion-\$8 billion; floods, \$2.41 billion; and hurricanes, \$1.2 billion-\$4.8 billion.

Loss of life from droughts in the United States is virtually nonexistent, although earlier accounts of this nation's history point to malnutrition deaths during some severe droughts (Warrick 1975). Contemporary drought losses are seen in withered crops, poorly hydrated livestock, and other human-use systems, such as recreation or water transport, that depend upon water.

Over the 1975-1998 study period, 5 years stand out as high-loss years (Figure 5-12). Note that the damage estimates reported in Figure 5-12 are much lower than those reported elsewhere, a consequence of our use of the estimates reported in *Storm Data*. According to these conservative estimates, droughts are responsible for a minimum of \$612 million in damages each year. Droughts in the Great Plains and western

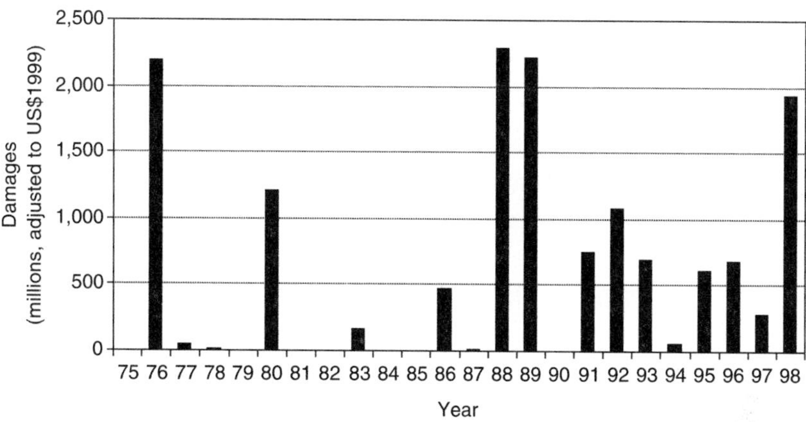

FIGURE 5-12 Trends in drought damages, 1975-1998 (in 1999 dollars).

United States in 1976 were responsible for a minimum of $2 billion in damages, although some estimates are as high as $15 billion (Riebsame et al. 1991). The National Climatic Data Center (NCDC 2000) reports $20 billion in damage from the 1980 droughts in the central and eastern portions of the country and another $40 billion from the 1987-1989 droughts, an estimate that makes it the costliest natural disaster during our study period. A southern drought stretching from the Carolinas to Texas generated an additional $6 billion-$9 billion in losses in 1998. Using these figures over our more conservative estimate would catapult drought into the top three hazards for economic losses, whereas our data place drought sixth. The greatest natural hazard loss collectively in the United States appears to revolve around water—either too much in the case of floods or not enough in the case of drought.

Extreme Heat

Drought and extreme heat are not synonymous. Irrespective of drought, approximately 200 deaths per year are attributable to heat stress in the United States (Kilbourne 1989). Linking deaths to heat stress as opposed to another cause, however, can be difficult. The 3 years with higher than normal casualties were 1980, 1986, and 1995 (Figure 5-13a). The 1980 fatalities numbered around 400 and may have been as high as 1,700 (FEMA 1997a) or possibly higher, depending on the data source (NCDC 2000). The 1995 heat wave in the Midwest contributed to the deaths of 670 people, mostly in urban areas such as Chicago, Illinois,

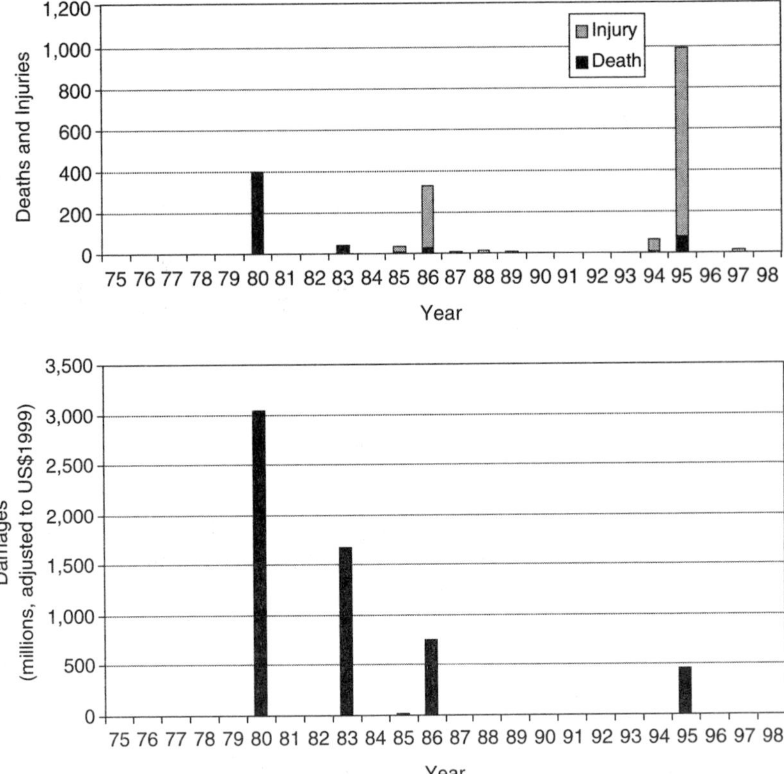

FIGURE 5-13 Trends in losses from extreme heat events, 1975-1998: (a) deaths and injuries, and (b) damages (in 1999 dollars).

and Milwaukee, Wisconsin. Extreme heat can also be responsible for other damages (Figure 5-13b). Livestock are sapped of their strength just as people are, while transportation infrastructure such as roads and railroads may warp under the intense heat. Most significantly, power failures, water shortages, and crop losses all contribute to calculation of losses from this hazard. The most damaging year (almost $455 million) for extreme heat was in 1985, which was an abnormally hot and lengthy summer for most of the nation.

Extreme Cold

As with drought and extreme heat, severe winter storms and extreme cold are related, yet distinct phenomena. A cold spell can occur without a severe winter storm and severe winter storms can plow through portions of the United States without the weather being severely cold. A cold snap at the wrong time in the growing season can ruin crops, even though not a flake of frozen precipitation falls. For instance, two freeze events hitting Florida in 1983 and 1985 together inflicted over $3 billion in losses to the citrus industry alone (NCDC 2000).

Cold weather caused a relatively large number of casualties in 1983, 1985, and in 1994 (Figure 5-14a) with more than 100 people reportedly dying as a result of exposure or hypothermia. Over the study period, annual fatalities and injuries averaged 10 and 17, respectively. Damages from severe cold peaked in 1989 and again in 1998 (Figure 5-14b). The annual damage average was about $119 million. The expansive power of frozen moisture contributes greatly to this total through the cracking of

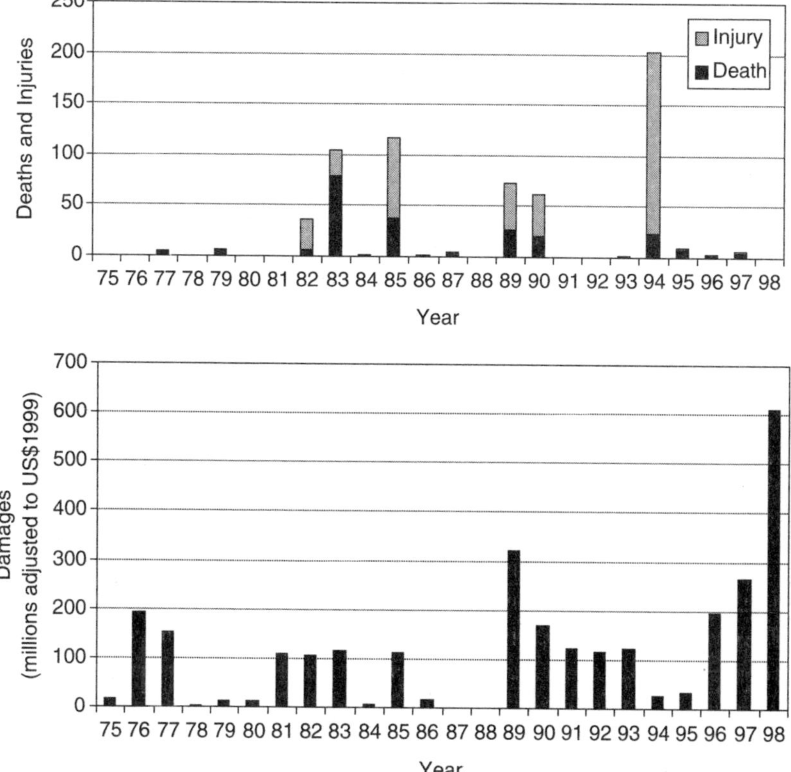

FIGURE 5-14 Trends in losses from extreme cold weather events, 1975-1998: (a) deaths and injuries, and (b) damages (in 1999 dollars).

roadways and building foundations, frozen power lines, and water-main breaks.

Severe Winter Storms

A large, concentrated amount of snowfall or ice can be very disruptive, particularly if the storm is long in duration. Communication and other lifeline infrastructure, such as transportation, are among the most heavily affected. Driving conditions often are unsafe and vast amounts of snowfall can cause roofs to collapse. Between 1975 and 1998, the annual number of winter hazard fatalities was 44 and injuries averaged 474 (Figure 5-15a). The peak years for fatalities were 1982 and 1983 with

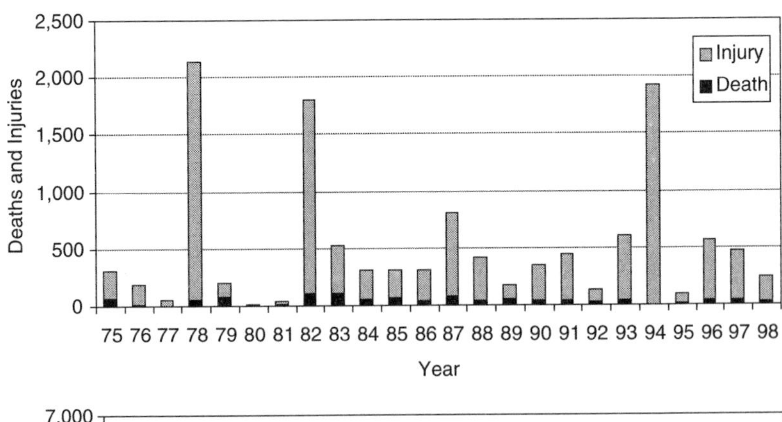

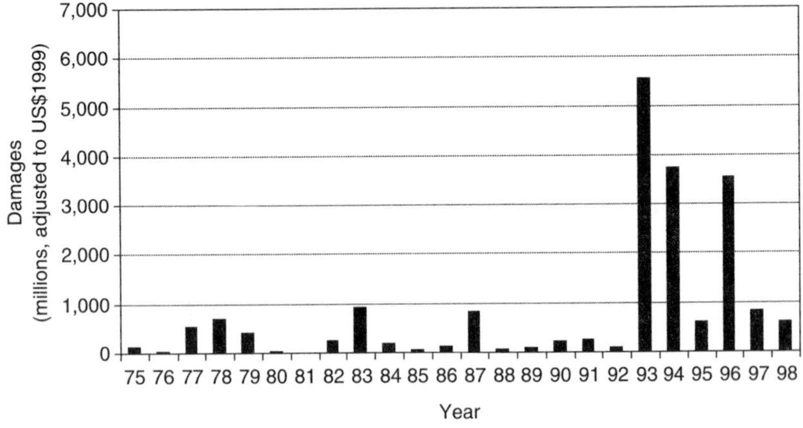

FIGURE 5-15 Trends in losses from winter hazards, 1975-1998: (a) deaths and injuries, and (b) damages (in 1999 dollars).

104 and 103 fatalities, respectively. The peak number of injuries (2,092) occurred in 1978, and is largely attributable to a blizzard that affected states from the Mid-Atlantic to New England. The death tolls vary for individual events, depending upon how the data compiler classified the death. Thus, the casualty figures reported here more than likely include incidences related to extreme cold, a category we have treated separately.

By far the three most damaging years for winter hazards were 1993, 1994, and 1996 (Figure 5-15b). With snowfall rates of 2 to 3 inches per hour, 1993's so-called "Storm of the Century" (March 12-14) paralyzed most of the eastern United States and inflicted between $3 billion and $6 billion in direct damages. Approximately 270 lives were lost and another 48 were reported missing at sea during this 3-day event (Lott 1993). The unusual Southeast Ice Storm of 1994 affected 11 states from Texas to Virginia (a large areal extent for an ice storm). At least nine deaths and $3 billion in damage was attributed to this event alone (Lott and Sittel ND). The Blizzard of 1996 dropped snowfalls ranging from 17 to 30 inches from Washington, D.C. to Boston, Massachusetts, as eastern seaboard commerce came to a standstill. Mail could not be delivered and federal government employees had an unexpected holiday. Approximately 187 deaths were attributed to the 1996 storm, with about $3 billion in damage/costs (NCDC 2000). Based on our data, the average damage total from 1975 to 1998 was $830 million per year. Despite the occasional big-event storm, winter hazards are a fairly pervasive threat and remain among the most difficult in terms of preparation and mitigation.

Hurricanes

Although not occurring as frequently as some hazards, hurricanes are among the most damaging—especially when making landfall in a heavily populated, unprepared place. Losses are caused by a number of factors including high winds, pounding rains, storm surge, inland flooding, and even tornadoes. Improved warning systems, building codes, and construction techniques have substantially decreased the number of deaths resulting from hurricanes in the United States compared to other parts of the world. Still, the continued development of vulnerable coastal areas puts more and more people and property at risk (Pielke and Landsea 1998).

The number and intensity of hurricanes during any single season is highly variable but does coincide with El Niño-Southern Oscillation

(ENSO) cycle, with large damage periods also associated with La Niña years (Pielke and Landsea 1999). Where one hurricane season sees relatively few, yet more intense, hurricanes (e.g. category 3 or higher on the Saffir-Simpson scale), the next season may have a greater frequency yet lower magnitude of storms (Figure 5-16a). The former pattern poses more danger because 83 percent of all damages are attributed to the 21 percent of landfalling tropical cyclones classified as intense (Categories 3-5) (Pielke and Landsea 1998).

Historically, the largest loss of life from any single hazard event in the United States resulted when a hurricane made landfall in Galveston, Texas, in 1900 (Larson 1999, Pielke and Pielke 1997). More than 8,000 people perished (Hebert et al. 1996). From 1975 to 1998 the average annual number of deaths caused by hurricanes was 16. Death tolls for individual storms may vary by source due to the timing of the fatality. For example, a death while repairing one's roof may not be attributed to the hurricane event, but a heart attack fatality during an evacuation might be. In terms of direct casualties (drowning, lightning, wind-driven events), the majority of hurricane fatalities during the past 25 years were people who drowned in freshwater floods caused by the excessive rainfall from the storm (Rappaport 2000). Interestingly, these are listed under hurricanes rather than floods, highlighting some of the classification difficulties with the loss data.

Five years—1978, 1985, 1989, 1994, and 1996—recorded at least 30 hurricane fatalities (Figure 5-16b). By far the greatest loss of life occurred in 1989 from Hurricane Hugo. Hugo, a category 4 hurricane, crossed the Virgin Islands and Puerto Rico before making landfall in South Carolina near the port city of Charleston. Fifty-seven lives were lost on the U.S. mainland; many more perished on the islands (NCDC 2000). Despite the 1989 peak, the loss of life from hurricanes in the United States has generally decreased over the past century.

Chronicling damage estimates tells another story. Annual losses average $3.1 billion for our study period. Four years exceed $5 billion in losses—1979, 1985, 1989, and 1992 (Figure 5-16c). The largest of these, 1992, exceeded $25 billion as a result of Hurricanes Andrew and Iniki. Ripping across Florida just south of Miami at the very beginning of the hurricane season, Andrew currently holds the record for causing the greatest amount of property damage ($27 billion) of any single natural event in the United States. Iniki added another $2.2 billion in losses. Surpassed only by Andrew, 1989's Hurricane Hugo wrought $9 billion in damage. The other loss peaks are attributable mainly to Hurricane

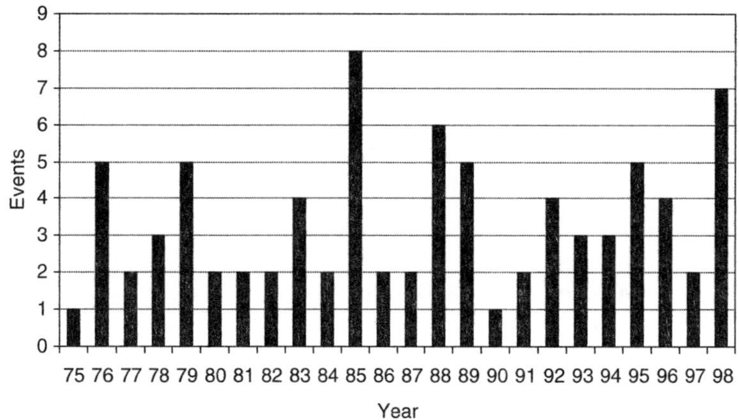

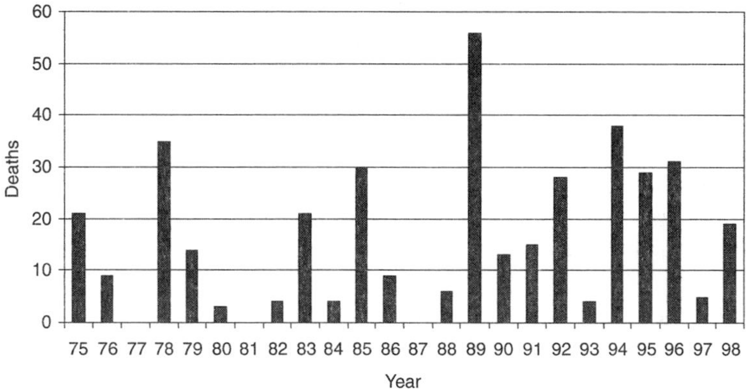

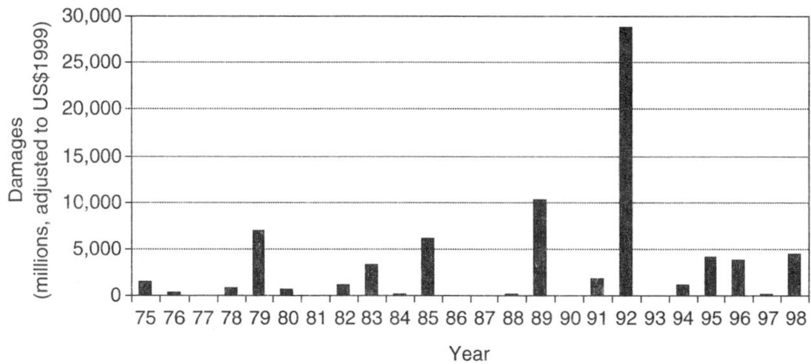

FIGURE 5-16 Trends in hurricane and tropical storm events and losses, 1975-1998: (a) events, (b) deaths, and (c) damages (in 1999 dollars).

Frederic (1979) and 1985's string of hurricane events that included Elena, Gloria, Juan, and Kate.

Hurricane damages will likely increase for two main reasons. First, the population growth along America's coasts is increasing. Second, it appears that the past 20 to 30 years have experienced a lull in "normal" hurricane activity. The number of annual events is expected to climb. Coupled with greater coastal development, damages will rise as well. Hurricane Floyd (1999) may be a harbinger of losses in the future. Insured losses already (in 2000) are in excess of $6 billion and climbing (Anonymous 2000, ISO 2000, NCDC 2000).

Earthquakes

Earthquakes result from the sudden release of accumulated stress within the Earth's crust. Usually occurring with little or no warning and with a rapid rate of onset, earthquakes have the ability to quickly cause extensive devastation and loss of life. Unlike weather-related events, there is no seasonality or periodicity to earthquakes. They vary considerably in their magnitude, and occur worldwide, with an average of more than 8,000 per day (NEIC 1999). Granted, most of these earthquakes are quite small and of minor consequence, being recorded by sensitive instruments, but not actually felt by people. Of foremost concern, however, are those few strong to great earthquakes that occur each year in areas where there are vulnerable human populations. Earthquakes also may trigger secondary threats such as landslides, tsunamis, and technological failure—threats to human settlements as well.

From 1975 to 1998, the number of earthquake epicenters within a state boundary ranged from a low of 7,800 (in 1975) to more than 70,000 (1992) (Figure 5-17a). Clearly, not all of these were felt earthquakes, let alone significant ones. The number of significant earthquakes (more than $1 million in damages, or ten or more deaths, or magnitude 7.5 or greater, or intensity X or greater), was quite varied as well, with a total of 45 during the 24-year study period. The greatest number of fatalities occurred in 1989 (Figure 5-17b), a year with only two significant events. One of these events was the devastating Loma Prieta earthquake that affected the San Francisco Bay/Santa Cruz area of California. With an epicenter located under a mountain of the same name near Santa Cruz, the Loma Prieta quake killed 62 people; many of these perished on the interstate highways (Figure 5-18). Striking near the height of a normal day's evening rush hour, the earthquake's death toll might have been

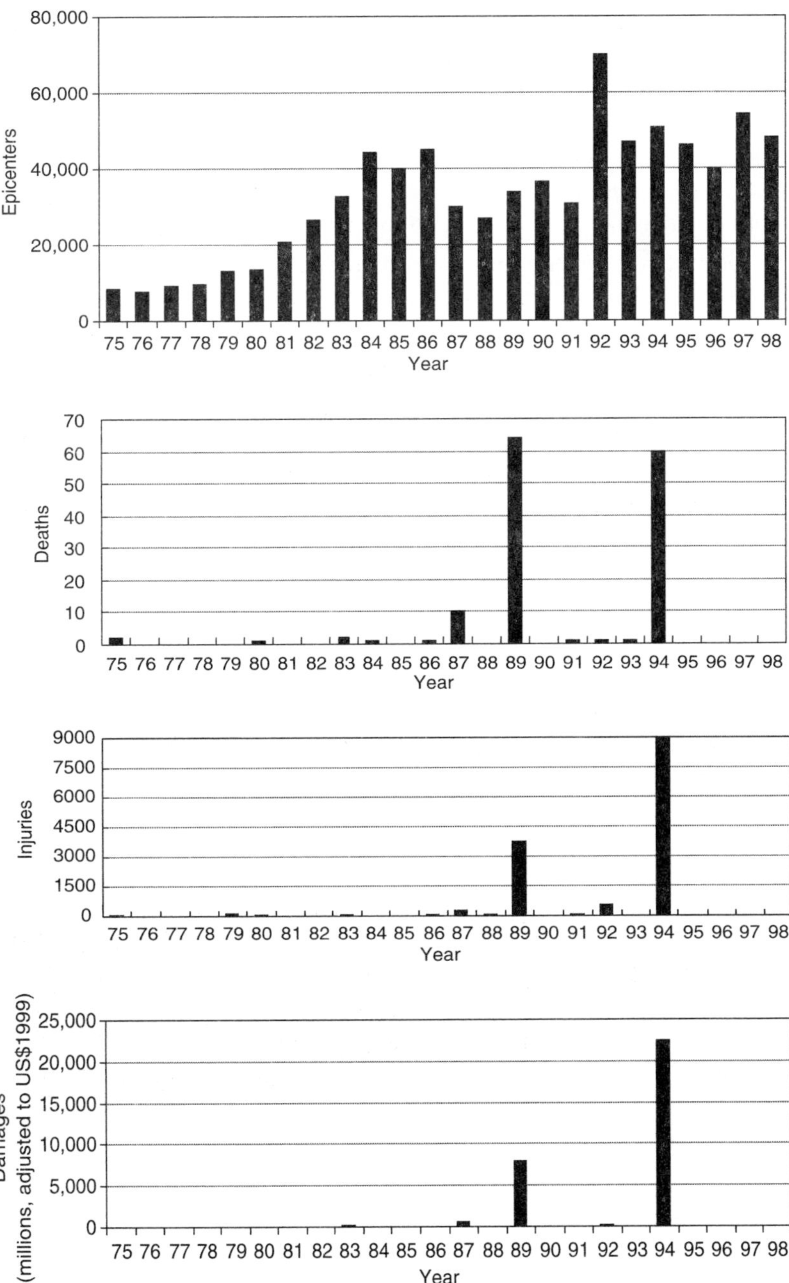

FIGURE 5-17 Trends in earthquake events and losses, 1975-1998: (a) epicenters within state boundaries, (b) deaths, (c) injuries, and (d) damages (in 1999 dollars).

even higher if not for a large number of people at home watching the locally played baseball World Series. Another 57 perished in 1994's Northridge quake to the northwest of Los Angeles, California, which recorded more injuries than the Loma Prieta earthquake (Figure 5-17c). A Richter magnitude earthquake of 6.7, the Northridge event occurred at 4:30 a.m. and killed many as they slept. Smaller losses were also seen in 1983's Coalinga and 1987's Whittier Narrows events.

Earthquake losses appear to escalate with each new major event (Figure 5-17d). Annual losses average approximately $1.3 billion. Loma Prieta's losses exceeded $5 billion, only to be topped by Northridge at over $20 billion, the costliest earthquake event in the continental United States. Other estimates attribute $358 million to Whittier Narrows and $6 billion to Loma Prieta (Palm 1995). In the absence of reliable earthquake prediction and warning systems, structural mitigation, land-use zoning, and insurance remain the best courses of action to limit or reduce the impact of future earthquake losses.

FIGURE 5-18 Damage to the Cypress Street Freeway extension in Oakland, California, from the 1989 Loma Prieta earthquake. Source: Photograph by Susan L. Cutter.

Volcanoes

Although the Hollywood imaginations that brought us films such as 1997's *Volcano* would have us believe otherwise, volcanic activity is geographically confined to a few select areas in the United States. However, no continuous loss database exists for this hazard. The primary threats from this hazard include lava and mudflows, floods from melted snow pack, super-heated gases, and airborne projectiles. Some data are available for the few major events that transpired between 1975 and 1998. The continuing eruption of Kilauea (Figure 5-19) in Hawaii (1983 and recurring episodes) has resulted in over $61 million in damages. Damage and loss of revenue from ash and debris flows from the Redoubt, Alaska volcano totaled about $160 million, making that 1989-1990 eruption the second costliest in U.S. history (NODAK 2000).

The most damaging volcanic event in our study period was the 1980 eruption of Mount St. Helens in the State of Washington. Losses to forestry, property, agriculture, income, transportation infrastructure, and the cost of cleanup from the eruption exceeded $1.5 billion (FEMA 1997a). More than 60 lives were lost. Many of these fatalities were backwoods campers and hikers and permanent residents who failed to heed

FIGURE 5-19 The Pu'u 'O'o cone in September 1983, early in the Pu'u 'O'o-Kupaianaha eruption of Kilauea volcano 1983-1986. Source: USGS Hawaiian Volcano Observatory Photo Gallery. *http://www.hvo.wr.usgs.gov/gallery*.

evacuation warnings. As with earthquakes, the most effective way to reduce losses from volcanic eruptions lies in land-use planning. Development of monitoring systems that detect physical changes within the environment may also provide crucial lead time for warnings and evacuations in response to an impending eruption.

Hazardous Material Spills

Up to now, this chapter has focused largely on "natural" hazards. Yet, the interaction between society, its technology, and natural systems also gives rise to a range of other hazards. Acute events, such as industrial accidents and oil or hazardous material spills, fall under this category, as do more chronic hazards, such as pollution. Dam failures can be considered as technological failures, but the effects of those failures (i.e., a flash flood) were treated earlier.

This section focuses on the losses caused by hazardous material accidents, although they represent only one of the many potential technological risks that face communities. These include releases by air, water, rail, and road carriers. Over the study period, hazardous material events peaked in the late 1970s, declined significantly through the 1980s, and had climbed again by the mid-1990s (Figure 5-20a). The decline in the number of events during the 1980s was a reflection of changes in reporting requirements in 1981 (spills of less than 5 gallons were no longer included) as well as improved safety (Cutter and Ji 1997). Although the occurrence of fatalities is more evenly distributed over time (Figure 5-20b), the temporal pattern for injuries mirrors that of the overall occurrences (Figure 5-20a). The average annual human loss was 24 deaths and 537 injuries, with the majority of fatalities from highway accidents.

Two peak years stand out for fatalities and injuries—1992 and 1996. Sixty-one deaths occurred in Wisconsin in 1992 and 111 occurred in Florida in 1996. The injuries in 1992 are the result of multiple events in several states, whereas a single event accounts for the large number of fatalities and injuries in 1996. A major railroad spill occurred near Alberton, Montana, in April 1996 when several cars went off the tracks, spilling chlorine and other materials and forcing hundreds of residents from their homes. Interstate 90 was closed for several days and 792 people were injured. During the study period, damages have clearly increased (from $20 million in 1975 to close to $50 million in 1998), with an average of $32 million per year (Figure 5-20c).

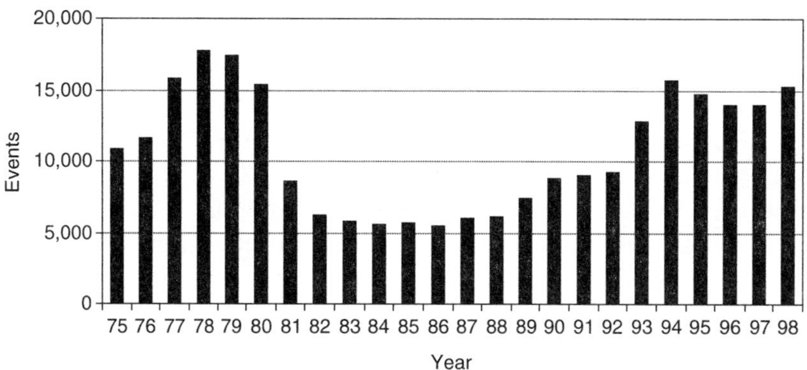

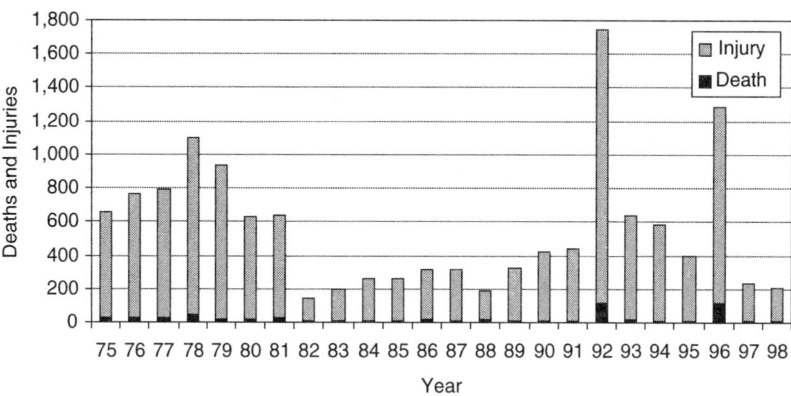

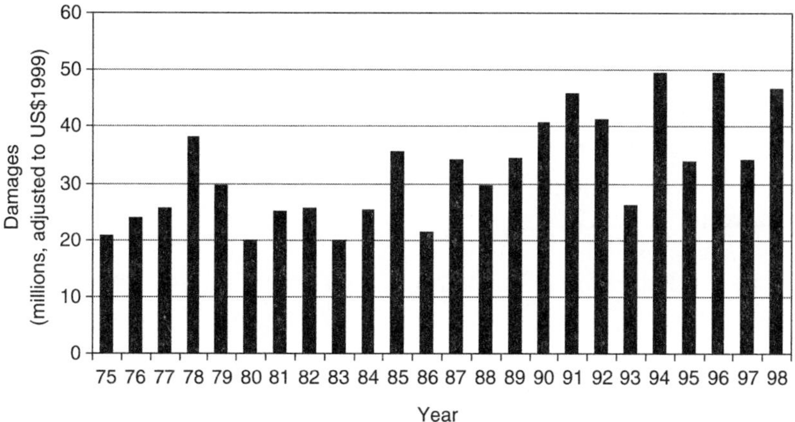

FIGURE 5-20 Trends in events and losses involving hazardous materials, 1975-1998: (a) events, (b) deaths and injuries, and (c) damages (in 1999 dollars).

CONCLUSION

Natural hazards conservatively cost this nation $12.5 billion annually during the past 24 years. Damages from natural hazards are highly variable from year to year and from decade to decade. However, steady increases in losses were found during the 1990s, the most disastrous decade ever. Fatalities remained relatively steady during the past 24 years.

Floods are the most costly natural hazard to this nation, in terms of both dollar losses and human fatalities. Hurricanes and tornadoes round out the top three hazards in overall economic losses according to our data. In fact, weather-related phenomena (floods, hurricanes, tornadoes, etc.) account for 89 percent of the total losses from natural hazards and 92 percent of the human fatalities. Geophysical events (earthquakes and volcanoes) account for 10.8 percent of the economic losses and 2 percent of the fatalities during this study period, whereas hazardous materials incidents account for less than 1 percent of the economic losses, but 6 percent of the total fatalities. Cumulatively, hazardous material incidents account for greater losses of life than earthquakes and volcanoes combined!

The hazard loss story for the United States over the past three decades is a mix of good and bad news. Save for those few major disaster events, injuries and fatalities for most threats have either declined in number or leveled out. Much of this can be attributed to the investments made in structural reinforcement of dwellings, warning systems, land-use planning, and education. A few singular disaster events contribute to the overall loss totals (Hurricanes Hugo, Andrew, and Iniki; the Northridge and Loma Prieta earthquakes; the Midwest floods). It is the cumulative impact of less catastrophic, yet more frequent, natural hazard events, however, that contributes to the escalator-like (up and down) trends in hazards events, losses, and casualties.

A burgeoning population and a desire to congregate in attractive, yet hazard-prone areas may alter these trends in the future, especially if we base our projections on the decade of the 1990s. For example, the population of all coastal counties has outpaced the total U.S. population growth by 15 percent in the past two decades (Ullmann 2000). Not only does this increase the level of exposure to enormous amounts of real property, but it increases the potential to erase the gains made in reducing hurricane fatalities and injuries during the past 24 years. Similar observations can be made regarding earthquakes and wildfires, or about a lifestyle and an economy that demands the production of hazardous materials.

This chapter has reported fatality, injury, and damage losses for 14 threats. It does not purport to be exhaustive; clearly, many other environmental threats exist and, more importantly, new threats and new losses continue to appear. Emotional, cultural, and ecological losses—although important and unignorable—have not been addressed in the data reported here. The loss estimates were derived from a variety of sources, are very conservative, and likely understate the severity of disaster losses to the nation. Without a centralized, national data collection effort focused on gathering identical physical, social, and spatial loss variables, loss reduction efforts will continue to be plagued by hazard incomparability, difficulty in conducting vulnerability assessments, and a poor understanding of the geographic dimensions of disaster.

CHAPTER 6

Which Are the Most Hazardous States?

Deborah S. K. Thomas and Jerry T. Mitchell

Hazard events and losses clearly vary geographically. For instance, the Gulf and Atlantic coasts of the United States are much more prone to tropical storms whereas the Pacific Coast is more seismically active than any region in the country. The juxtaposition of where people live and work and hazard-prone areas increases the potential for loss of life and property damage. It is equally important to understand the geographic patterns of hazard events as well as the spatial distribution of human occupance of hazardous areas. In this way we can begin to assess where the risk and vulnerability are greatest and why.

This chapter examines regional patterns of hazard events and losses for the nation. The data sets utilized in this chapter are the same as those described previously. Consequently, the same limitations apply to the interpretation of the spatial patterns as the temporal trends (Chapter 5). Specifically, all damage categories are assigned a dollar amount based on the lowest value of the range given, and thus are extremely conservative estimates. Further, all damages are adjusted to 1999 U.S. dollars, in tables as well as maps unless otherwise stated. The maps use the natural-break method for classifying data, resulting in five categories ranging from low to high. This chapter begins by presenting cumulative damage and casualty statistics by state, followed by a discussion of the geographic distribution of losses from individual hazards. It concludes with a regional summary of damaging events.

GEOGRAPHIC SCALE AND LOSS INFORMATION

Geographic scale is a limiting factor for almost all of the databases utilized. Meaningful loss data often are lacking at the local level, and attributing the loss data to a specific geographic unit, such as a county or community, is not always possible. For example, the Storm Prediction Center collects very detailed tornado event information that includes location, path, deaths, injuries, and a damage category. However, a single tornado event may have affected more than one political jurisdiction, complicating the assignment of losses to specific counties. On the other hand, if one were interested in very detailed information about a single tornado event, these data are quite useful. This scale limitation is true of most of the event-based data sets, including earthquakes, hurricanes, and meteorological events listed in the National Weather Service's (NWS) *Storm Data and Unusual Weather Phenomena*. Finally, some of the hazards data are not even collected or reported below the state level. Historic losses associated with hazardous materials transportation spills are an example of this type of geographic limitation in the data during our study period.

LOSSES FROM ALL HAZARD TYPES

Of the nearly 9,000 deaths caused by hazard events in the United States, most occurred in Texas, the Southeast, and the eastern Great Lakes regions (Figure 6-1a). The Northwest, New England, and the upper Great Plains experienced significantly fewer fatalities. Texas alone accounted for 10 percent of all deaths, or more than five times the national fatality average and nearly twice the next highest state (Table 6-1). California and Colorado also stand out in the geographic distribution of hazard fatalities. Although not the only state with many fatalities from hazard events, Colorado's high ranking is attributed to the Big Thompson flood of 1976, when at least 139 people were killed. Deaths in California were caused predominately by three hazards: earthquakes, flooding, and winter-related storm events. The Loma Prieta and Northridge earthquakes account for over 25 percent of fatalities for the state during this time period. In Texas, the fatalities were attributed to a broad

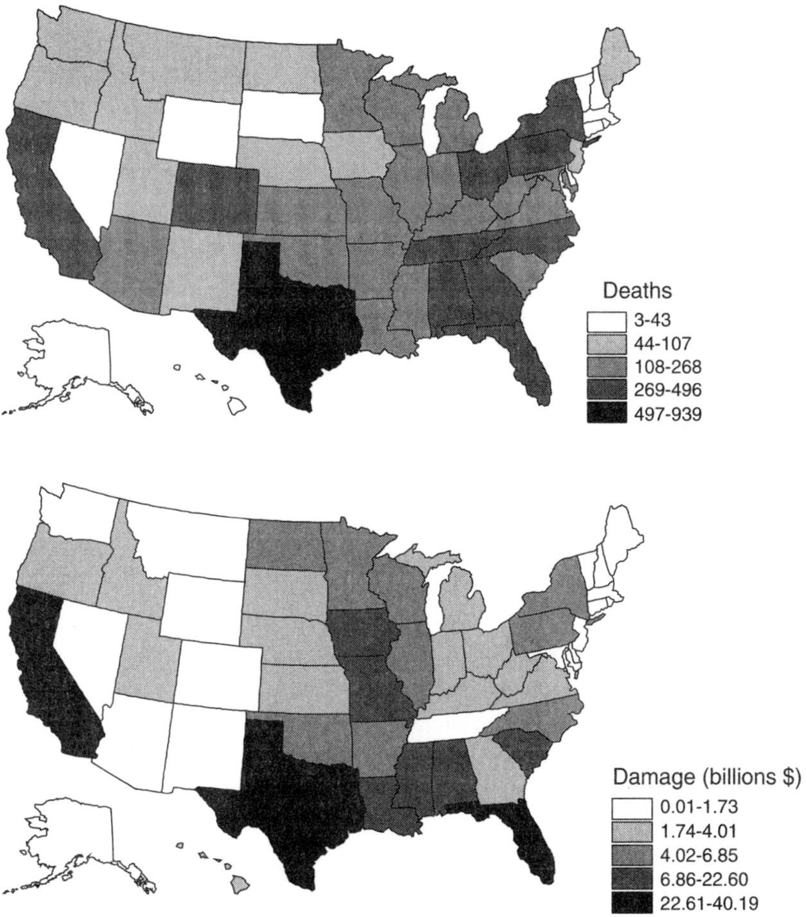

FIGURE 6-1 Geographic patterns of loss from all hazards, 1975-1998: (a) number of deaths, and (b) damages (in 1999 dollars).

range of hazards—hurricanes, floods, and tornadoes. Hazardous material spills and lightning caused nearly 70 percent of the listed fatalities in Florida. Florida also leads the country in the number of lightning deaths by a factor of two.

The geographic distribution of damages from all hazard types shows a slightly different pattern than the fatality maps (Figure 6-1b). The central portion of the United States bordering the Mississippi River incurred substantial amounts of damage, as did the Gulf Coast states, the

TABLE 6-1 Top 10 State Death Totals from All Hazards, 1975-1998

State	Deaths
Texas	938
Florida	495
Alabama	440
California	415
New York	410
Tennessee	367
Pennsylvania	347
Georgia	333
Colorado	294
North Carolina	291
National average	180.0

Carolinas, Florida, and California. California lives up to its reputation as the "disaster state"; not only does it rank fourth in number of deaths caused by hazards, it also endured the largest total losses at over $40 billion (Table 6-2). In California's case, a majority of the damages were from earthquakes and flooding. The losses in Texas stemmed from hurricanes, droughts, floods, and tornadoes, which together caused over $20 billion in damages. Not surprisingly, given Hurricane Andrew's ranking as one of the nation's most damaging natural disasters to date, the majority of damages in Florida resulted from hurricanes, although floods and tornadoes contributed over $1 billion themselves.

TABLE 6-2 Top 10 State Damage Totals from All Hazards, 1975-1998

State	Damages (billions, adjusted to U.S. $1999)
California	40.1
Florida	33.0
Texas	22.0
Iowa	15.6
Louisiana	12.4
Mississippi	12.2
South Carolina	12.1
Alabama	11.7
Missouri	11.4
North Dakota	6.8
National average	6.0

SPATIAL VARIATION IN HAZARD EVENTS AND LOSSES

Using these aggregate numbers, Texas may be viewed as the "fatality state" and California the "disaster state." When individual hazards are examined over the past 24 years, a very different geography of hazards emerges. The spatial variation in events, injuries and deaths, and damage totals highlight the differences between catastrophic events and everyday occurrences of hazards—both of which contribute to America's hazardscape. For the ranking of the top five states for hazard events, deaths, and economic damages for each individual hazard see Appendix B.

Floods

Very few places are completely protected from flooding. In fact, every state had at least four deaths from this hazard and a minimum of $10 million in property and crop loss during our study period. Most flood-related deaths were concentrated in the southern tier of the United States extending up through the Midwest and into New York and Pennsylvania (Figure 6-2a). Pennsylvania's place at number 2 is a direct consequence of flooding in Johnstown that killed over 75 people in 1977. Interestingly, this was not the first experience with serious flooding by this community. Flooding in 1889 in Johnstown holds the record as the second-deadliest disaster in U.S. history when over 2,200 people died (FEMA 1997a). Colorado is in the top five states as a direct consequence of the 1976 Big Thompson flood that killed more than 139 people, an event that focused national attention on the threat of flash floods (Gruntfest 1996).

Large dollar damages are concentrated in four states: three that border the Mississippi River (Iowa, Missouri, and Louisiana) and Texas (Figure 6-2b). Texas tops the charts for both flooding deaths and damages. Of the 42 events listed as causing more than a billion dollars in damages between 1980 and 1998, 4 were Texas floods (NCDC 2000). These include flooding in southeast Texas in fall 1998, Dallas in May 1995, the Harris County flood in October 1994, and the May 1990 floods in north central and east Texas. The great Mississippi River flood events of 1993 and 1995 are clearly evident on the map of flood damages especially Iowa and Missouri. States throughout the central portion of the United States were severely impacted, causing in excess of $20 billion according to some estimates (Changnon 1996). This 500-year flood event raised serious questions about the prudence of strict river control

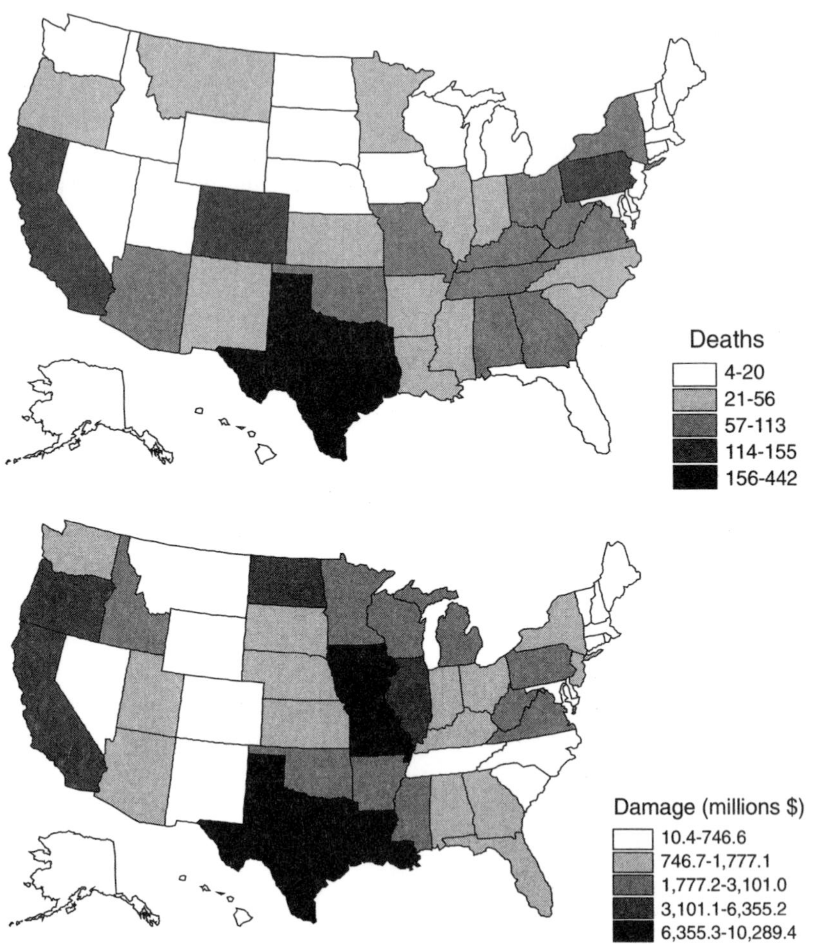

FIGURE 6-2 Trends in losses from flooding, 1975-1998: (a) number of deaths, and (b) damages (in 1999 dollars).

practices that officials have routinely implemented to protect people and property from flooding in the region (IFMRC 1994). However, flooding along the Mississippi and its tributaries is very much a part of the region's cultural and economic history (Barry 1997, Changnon 1998).

Flooding has also plagued the northern Great Plains where the Red River has repeatedly overflowed its channel (IJC 2000). The impacts were so widely felt in Grand Forks, North Dakota, and in East Grand Forks,

Minnesota, after the 1997 flooding, that the Federal Emergency Management Agency (FEMA) and the North Dakota Division of Emergency Management started the mass distribution of *Recovery Times* in April as a resource for the those in the community (FEMA 1997b). Flooding is also a perennial problem in California, where rains combine with snowmelt from the mountains to cause severe winter and spring floods. Among the most notable of these were the 1997 floods involving the Sacramento, Mokelumne, San Joaquin, and Tuolumne rivers in California's Central Valley.

Tornadoes

Although the areal extent of damage from this hazard is limited when compared to other events such as hurricanes, the damage within these small areas is often catastrophic. Most places within the United States are susceptible to tornadic activity, but some areas have greater exposure than others. States in the south-central region (Kansas, Oklahoma, and Texas, in particular) have the highest annual average of tornadoes (Figure 6-3a), leading to the regional moniker of "Tornado Alley." This concentrated pattern of tornadoes is a result of atmospheric instability caused by the collision of colder northern air masses with warm, moist air from the Gulf of Mexico, especially during late spring and early summer. Florida also ranks near the top in overall tornado events.

Tornadoes are the leading cause of all injuries resulting from natural hazards. There is a slight regional shift in the spatial pattern of death, injuries, and damage away from the event core in the south-central Great Plains to the southeast (Figures 6-3b and 6-3c). In other words, although Tornado Alley clearly has the most tornado events, these areas do not necessarily receive the greatest level of damage or casualties (deaths and injuries). The rankings of top states (Appendix B) reinforce this point, particularly where deaths are concerned. This may be a result of different housing types, greater population concentrations, time of day, or even lack of warning systems in the southeastern region.

Economic damages are concentrated in the Great Plains and southern states as well (Figure 6-3c). Texas ranks first for both deaths and damage from tornadoes, with almost twice as many as the next closest ranking state. The Red River tornado outbreak of April 10, 1979, illustrates just how devastating Texas tornado history has been. A single tornado, at times 1.5 miles wide, ripped through Archer, Wichita, and Clay counties, killing more than 50 people, injuring almost 2,000, and

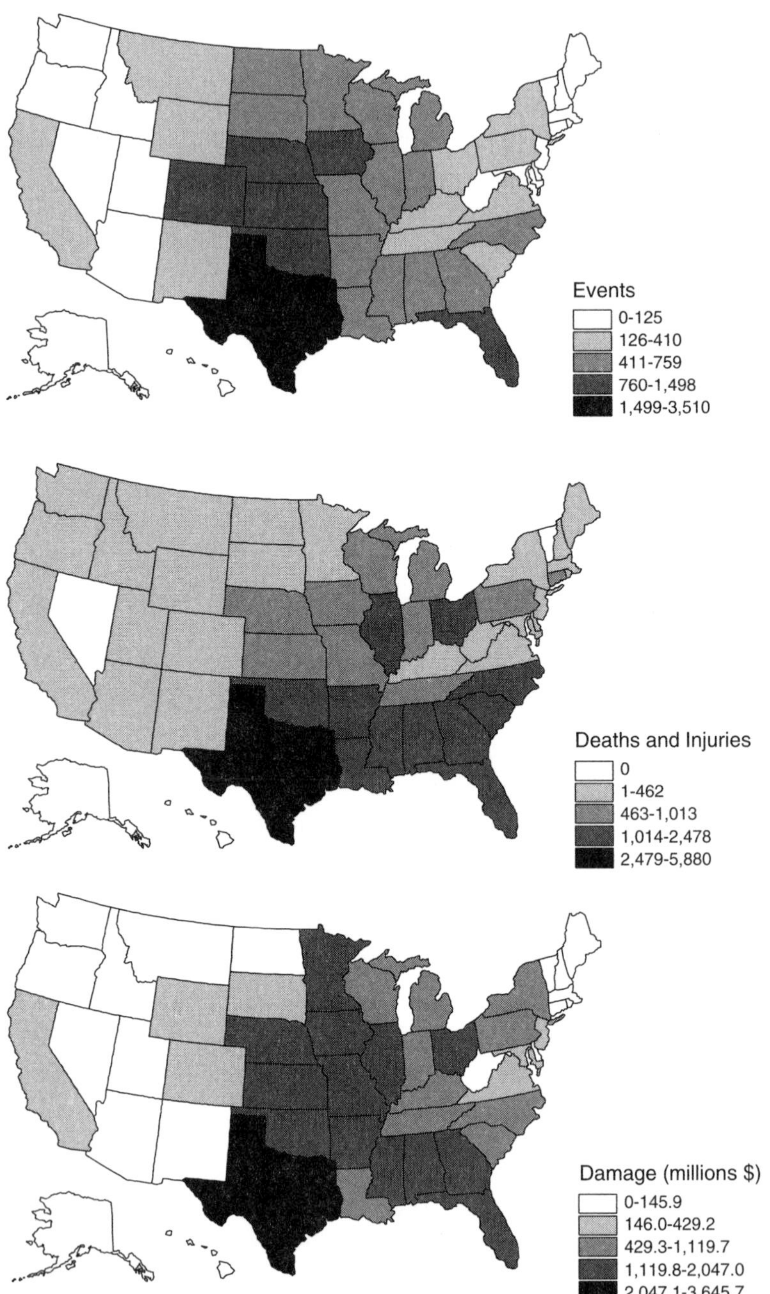

FIGURE 6-3 Trends in tornado events and losses, 1975-1998: (a) number of events, (b) deaths and injuries, and (c) damages (in 1999 dollars).

causing around $400 million in property damage (NCDC 1979, Bomar 1995). Along with its location in an area susceptible to severe thunderstorms in the spring and summer, the sheer size of the state when compared to others is a factor in mapping these patterns.

Hail

Although no place is immune from hailstorms, the Great Plains experience the greatest number of annual occurrences. In fact, when mapped, hail has a more concentrated geographic pattern compared to most other hazards (Figure 6-4a). The spatial distribution of loss is equally concentrated (Figures 6-4b and 6-4c). Of the more than 100,000 hail events during the 24-year study period, only about 2 percent resulted in damages greater than $50,000. However, when intense hail events do occur, automobiles and crops in particular show the signs of damage. The top states for hail damages are concentrated in the Great Plains region. Similar to the pattern of tornadoes, Texas is number 1 for events as well as damages from hail, again partly due to climate, location, and its sheer size. Of all the hazards reported in this chapter, except wildfire (which is grossly underestimated), hail caused the fewest number of deaths (15) during our study period.

Thunderstorm Wind

As with hail and tornadoes, thunderstorm wind events are closely linked to severe thunderstorms. Admittedly, downslope winds, squall lines, tornadoes, and tropical storms can all produce damaging winds. Tornadoes and hurricanes were treated separately, partly because of their unique characteristics and partially because wind events are defined in *Storm Data* as thunderstorm winds near or in excess of 50 knots. As a result, regions more prone to thunderstorms also will experience a greater number of thunderstorm wind events. Wind events occur more frequently in the eastern half of the United States (Figure 6-5a).

The pattern of casualties highlights the southern tier of states as well as those along the Great Lakes (Figure 6-5b). The West and New England have very few thunderstorm wind casualties. As was the case with hail and tornadoes, Texas had the largest number of wind events, deaths, and damages. New York and Michigan both ranked high for deaths and damages (Figure 6-5c), mostly attributed to two specific thunderstorm events. In 1991, thunderstorms moved through the Michigan counties of

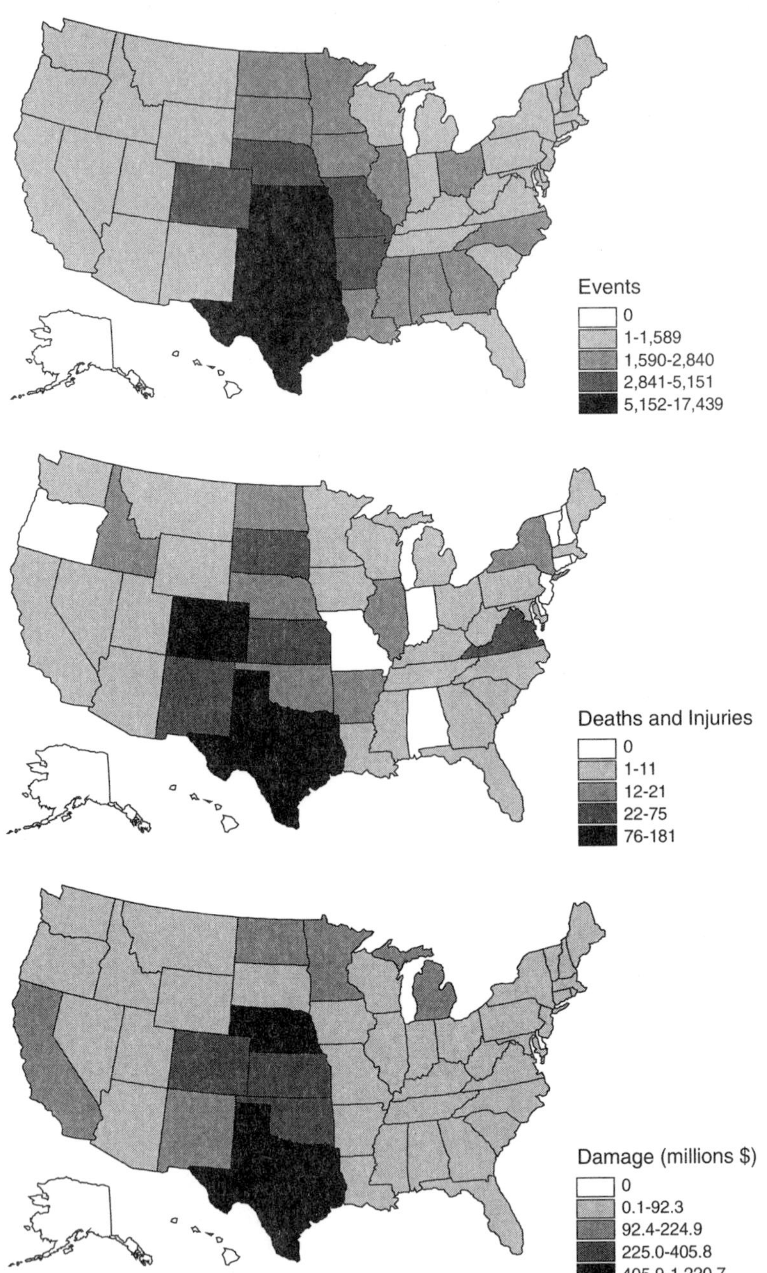

FIGURE 6-4 Trends in hail events and losses, 1975-1998: (a) number of hail events, (b) deaths and injuries, and (c) damages (in 1999 dollars).

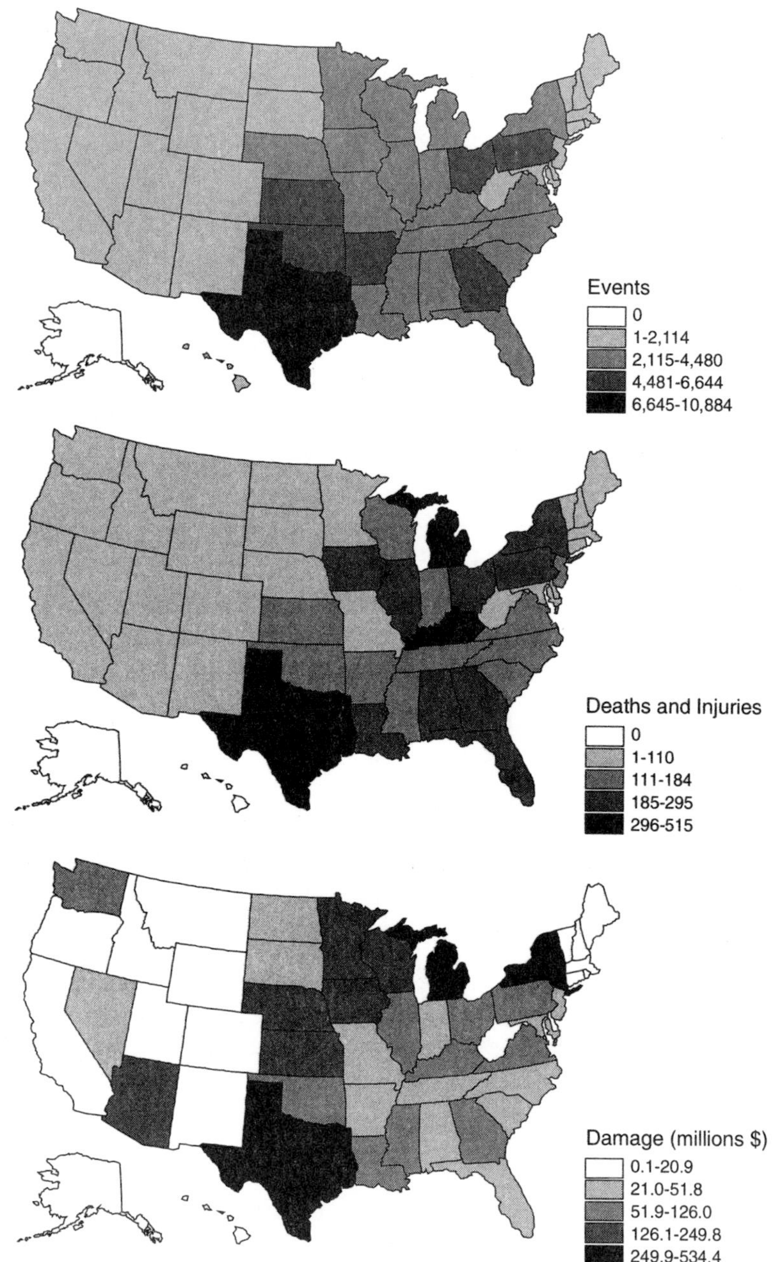

FIGURE 6-5 Trends in thunderstorm wind events and losses, 1975-1998: (a) number of thunderstorm wind events, (b) deaths and injuries, and (c) damages (in 1999 dollars).

Washtenaw, Ann Arbor, and St. Clair, causing over $20 million in damages and leaving more than 100,000 people without electricity (NCDC 1991). Eastern New York experienced one of the most intense and severe thunderstorm outbreaks of the century in May 1995 when 900,000 acres of forestland and 125,000 acres of commercial timber were damaged and over 200,000 people were left without electricity (NCDC 1995). Only a few days before, this same storm system destroyed 100 miles of phone and power lines in Michigan.

Because of the similar initiating conditions—collision of air masses producing severe thunderstorms, it is often difficult to differentiate and therefore classify tornado, hail, and thunderstorm events. For example, a tornado may be accompanied or preceded by hail. Is this listed as a hail event or a tornado? Similarly, microbursts may create substantial damage, yet unless there is other confirming evidence (tornado spotter, Doppler radar signal), it would be listed as a wind event. Needless to say, there are ambiguities in the exactness of the data used to map these meteorological hazards, and so, caution should be exercised in interpreting the findings.

Lightning

Of all natural hazards, lightning is one of the most ubiquitous and deadly hazards. Over 20 million cloud-to-ground flashes are detected every year in the continental United States (NOAA 2000b). According to our estimates, lightning ranks second after flooding in fatalities during the 24-year period, the majority of these in Florida. On the basis of a longer time frame (35-year record), others suggest that lightning is the leading cause of death from all natural hazards, even surpassing flooding (Curran et al. 1997). According to Curran et al., over 70 percent of deaths occur in the summer months in the late afternoon, where more than 80 percent of the fatalities are males. The seasonality and time of day are not surprising, given that thunderstorms often develop in the late afternoon and people are outdoors in warmer weather. Golf and other similar sporting events along with outdoor occupations put a large number of people at risk. The reasons for the disparate numbers for men and women are not easily explained and may be reflective of behavioral differences in activities between these two groups, including a predilection toward risk taking. Demographic differences such as age (young/old), gender (male/female), or race (Black/White) are rarely recorded in most hazards databases. The database on lightning fatalities is the only hazards database that we found that systematically records gender, for example.

The geographic distribution of casualties is concentrated in the Great Lakes region, the South, and the southern Great Plains (Figure 6-6a). The spatial pattern of damage is different with distinct clusters in the northwest, south-central, southeast, and Great Lakes states (Figure 6-6b). Oregon appears high as a result of lightning-caused wildfires, where losses were attributed to this hazard rather than wildfire. Overall damage estimates from lightning are low compared to other hazards. Even though the direct economic losses appear low from our data, the Na-

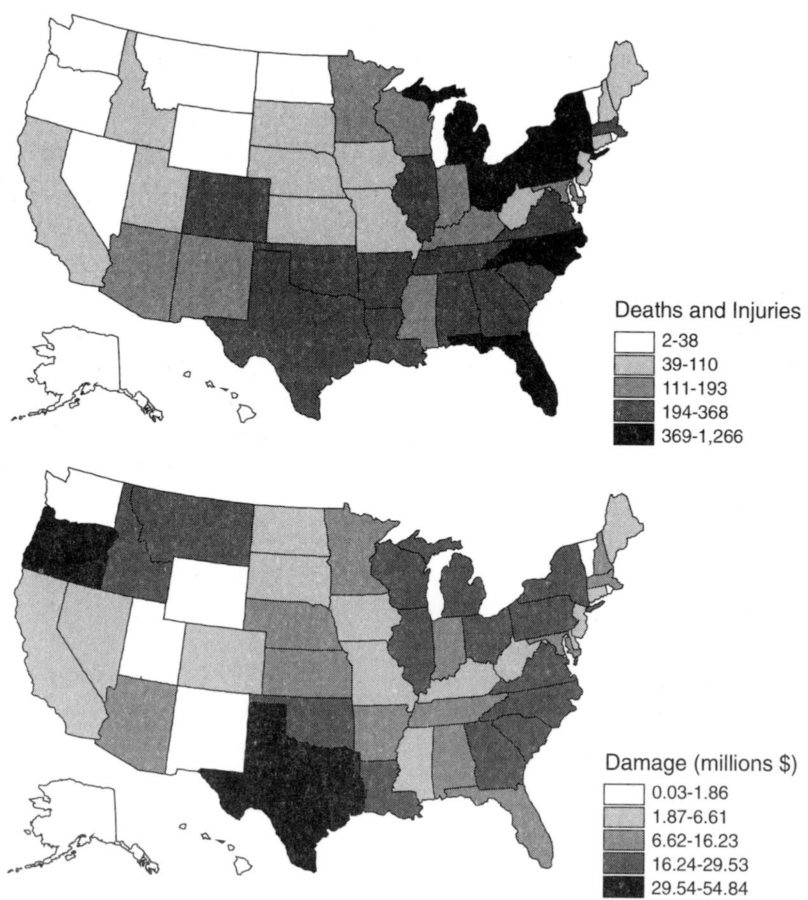

FIGURE 6-6 Trends in losses from lightning, 1975-1998: (a) deaths and injuries, and (b) damages (in 1999 dollars).

tional Lightning Safety Institute (NLSI 2000) suggests that, in reality, damages from lightning are much higher. It causes substantial damage when forest fires are started, houses are burned, airplanes are struck, and when power outages and surges ruin electronics. For example, the damages associated with the western fires in the summer of 2000 (caused by lightning) exceeded $1 billion.

Wildfires

Wildfires are a natural part of ecosystems, and human land use has increased the impacts of wildfires on society. Not all wildfires are caused by natural sources (e.g., lightning strikes). Most fires are caused by people, yet the average acreage burned is higher for those started by lightning (NIFC 2000). The urban-rural interface poses a particular challenge to wildfire management. As people move into areas that infringe on natural ecosystems or where building (or certain building materials) may not be appropriate because of the fire potential, danger to people and property increases. The Oakland, California, fires in 1991 catapulted the state to the top ranking for wildfire damage during our study period. This singular event revealed just how volatile that interface can be when climate and weather conditions are right; approximately 25 people died and nearly $1 billion in property was lost in this event. In 1993, fires in Malibu, California, caused over $500 million in losses (Figure 6-7). Most of the exclusive homes that burned were insured, and so, this particular disaster was somewhat unique because those individuals primarily impacted were a very select set of people who could absorb the cost of the property damage (Davis 1998).

Damage from wildfires is concentrated in the western region of the United States (Figure 6-8). A generally dry climate in this part of the country creates ideal fire conditions. Florida also experienced several severe wildfires during the study period. In 1998, for example, over 2,000 wildfires burned almost 500,000 acres, causing millions of dollars in damage (NIFC 2000). The ever-expanding urban fringe in this state along with drought conditions produced extensive damage potential, widespread burned areas, and poor air quality during the fires.

In terms of loss of life, few deaths were reported in the data used here. However, Mangan (1999) reports that 33 people died while fighting wildfires during the 1990-1998 period alone, and many more were injured. Because of the limitations in the data, our figures represent a significant underestimate of loss from this particular hazard.

FIGURE 6-7 Wildfire damage to homes in Malibu, California. Source: Photograph taken by Sally Cutter.

Extreme Heat

Even though not reflected entirely in the fatality statistics derived from *Storm Data*, exposure to severe heat is one of the deadliest of all natural disasters in the United States, killing between 150 and 1,700

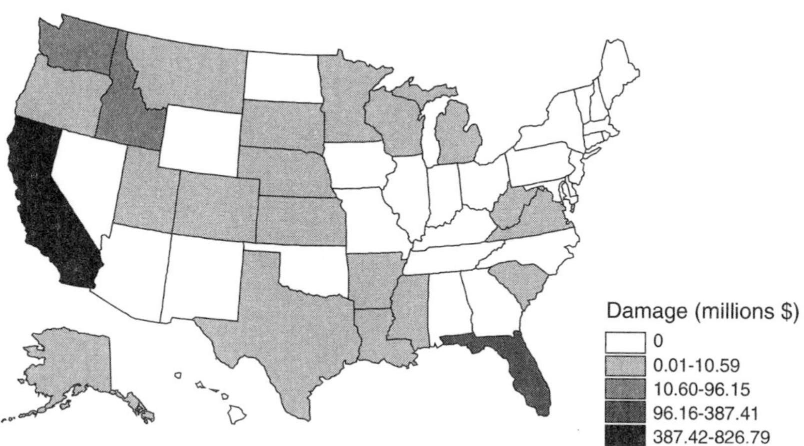

FIGURE 6-8 Wildfire damages, 1975-1998 (in 1999 dollars).

people each year (CDC 1995). In fact, one event in 1980 in Chicago caused 1,700 fatalities. Heat waves in 1983, 1988, and 1995 resulted in 1,500 deaths (NOAA 1995). These estimates suggest that exposure to extreme heat potentially kills more people than hurricanes, floods, tornadoes, and lightning combined, even though the databases we utilized do not indicate this.

A majority of deaths and injuries from extreme heat occurred in the East generally and in a few southern states, and extended up through the upper Midwest specifically (Figure 6-9a). Note that the Southwest and

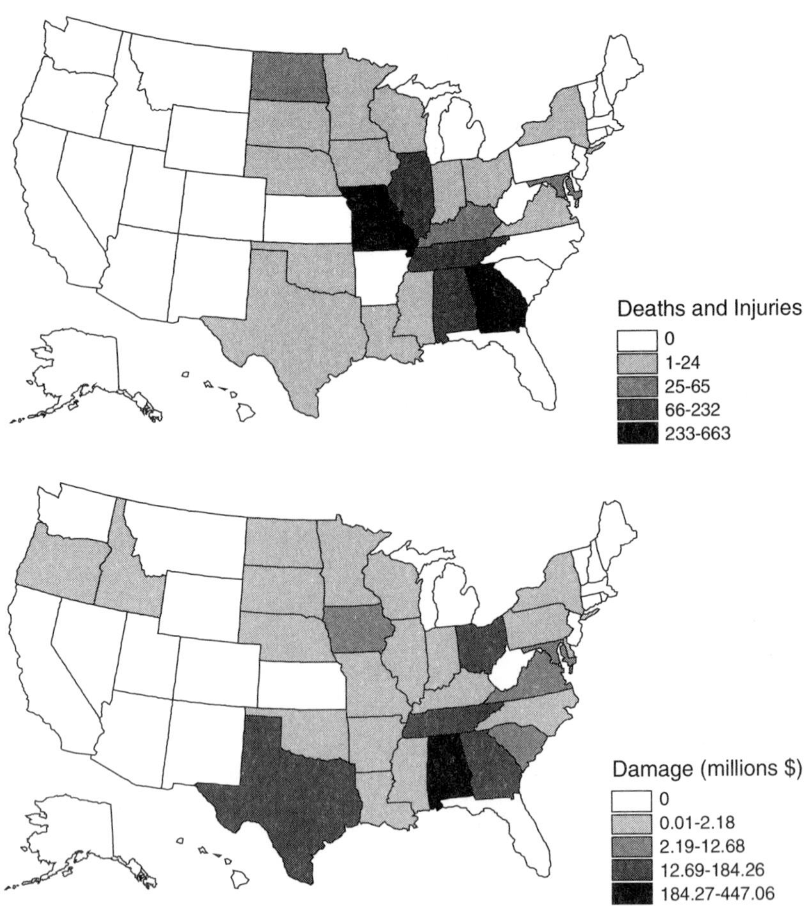

FIGURE 6-9 Trends in losses from severe heat events, 1975-1998: (a) deaths and injuries, and (b) damages (in 1999 dollars).

California have very few injuries and deaths from extreme heat, despite their warm climates. Texas ranks among the top five states for economic damages, experiencing over $100 million in losses from 1975 through 1998, but Alabama tops the list (Appendix B). Damages from extreme heat tend to be concentrated in the South and the Midwest as well and primarily reflect agricultural losses (crops and farm animals) (Figure 6-9b).

Drought

Although not precisely the same hazards, drought and extreme heat are intertwined and often misclassified. The difference between an event being recorded as heat rather than drought simply is attributed to the way states classify climate hazards when reporting to *Storm Data*. Drought by itself is an unlikely candidate for directly causing human fatalities as long as there is a supply of drinking water and food. In fact, within *Storm Data*, few deaths are attributed to drought. Both drought and extreme heat, however, display a similar geographic pattern of economic loss (Figures 6-9b and 6-10). Drought losses are mostly confined to the eastern Great Plains states, the South, and the Southeast. Iowa has the most extensive (in dollar amounts) damage totals, with over $3 billion during the study period.

The dollar losses associated with drought are underestimated in *Storm Data* and thus in our data as well because of its long-term impacts

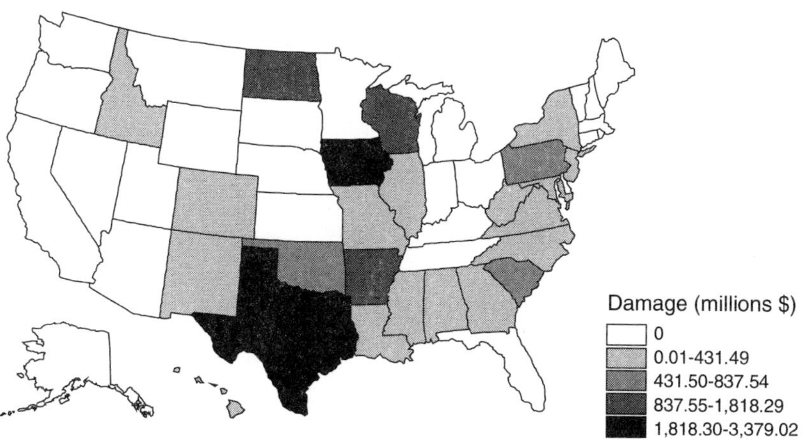

FIGURE 6-10 Drought damages, 1975-1998 (in 1999 dollars).

and difficulty in measurement. Other data sources provide greater loss estimates for this hazard. For example, of the 42 billion-dollar weather disasters during our study period (1975-1998) six were droughts. According to the National Climatic Data Center, these six events caused more than $74 billion in damages and 15,000 fatalities (NCDC 2000). Drought is the leading hazard in terms of total losses and in the most losses attributed to a single event, the 1988-1989 Midwest drought tallied almost $40 billion. With climate becoming more variable, long-term drought may be the most significant hazard facing this country in the future.

Extreme Cold and Severe Winter Storms

Memories of waking up to a white-covered ground and listening to the endless list of school cancellations in anticipation of hearing one's school name called are vivid for many people, but, what happens when parents cannot get to work, either because roads are actually too dangerous or because they must take care of their children? Severe winter storms and extreme cold have the potential to disrupt vast numbers of people's lives, often unexpectedly. For example, a major storm blanketed most of the East Coast in 1996—effectively closing federal government offices in Washington, D.C., for 2 days.

As the map of winter storm damage shows, nearly all portions of the United States except Hawaii have some loss experience with snowstorms, ice storms, or blizzards (Figure 6-11a and Figure 6-11b). Deaths are concentrated in New York, which has three times as many fatalities as the next highest state, California. With the exception of New York, winter storm deaths were more likely to occur in states that have slightly warmer climates than that of the Northeast. Perhaps somewhat unexpectedly, the two top-ranking states for winter storm damage are located in the Deep South (Alabama and Mississippi) with over $5 billion in losses each for the 24-year study period.

Damage and casualties from extreme cold were much less widespread, geographically, than winter storms (Figure 6-12a and Figure 6-12b). The Carolinas and Louisiana had around 90 deaths related to extreme cold, perhaps because people in these states are not as acclimatized to the cold or the houses are not well insulated against cold temperatures. The homeless, poor, and indigent also are more susceptible to the effects of cold weather, further increasing their vulnerability to this hazard. Pennsylvania ranks high as well perhaps due to urban poor popu-

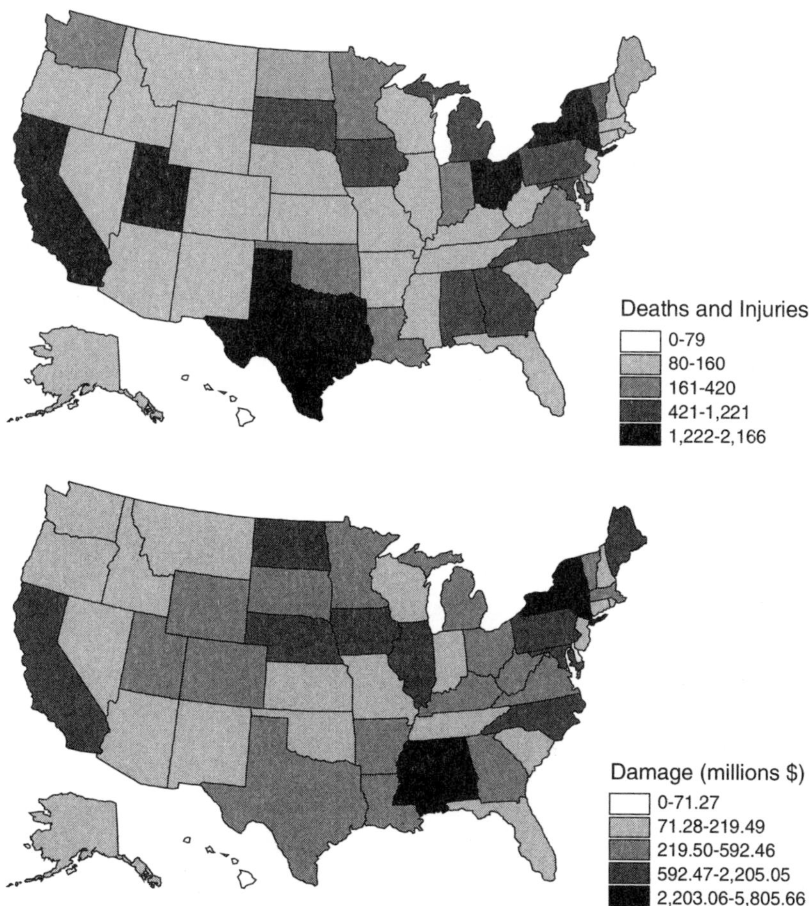

FIGURE 6-11 Winter hazards, 1975-1998, based on events causing more than \$50,000 in damages: (a) deaths and injuries, and (b) total damages (in 1999 dollars).

lations who cannot afford high heating bills during the cold season. The greatest amount of economic damage from extreme cold events occurred in Florida and California, primarily from freezes affecting agriculture. Although significantly underestimated in dollar amounts (when using *Storm Data*), Florida's citrus crops are extremely vulnerable to freezes.

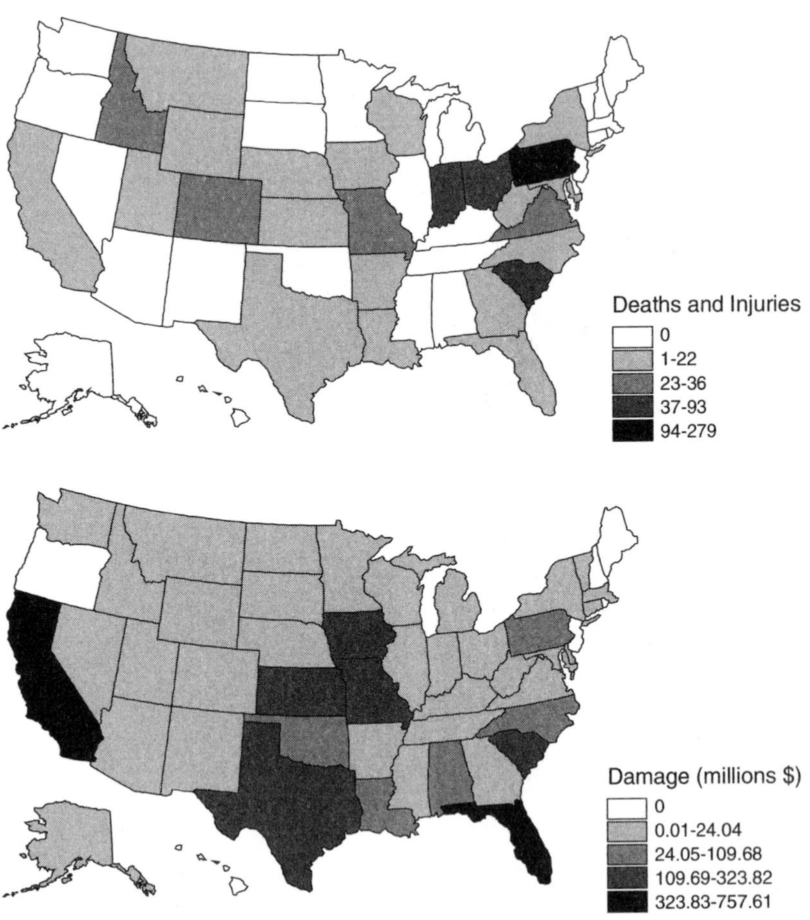

FIGURE 6-12 Extreme cold events, 1975-1998: (a) deaths and injuries, and (b) damages (in 1999 dollars).

Two different freeze events in 1983 and 1985 caused over $3 billion in damage, for example (NCDC 2000).

Hurricanes and Tropical Storms

The Atlantic and Gulf coasts of the United States experience hurricanes more frequently than the western United States except Hawaii. During the 1975-1998 time period, 72 tropical storms or hurricanes crossed eastern or Gulf Coast states, while 10 Pacific Ocean hurricanes made landfall in the United States (4 in Hawaii). The storms were not always hurricane strength when they made landfall; they may have weakened to tropical storms or even tropical depressions. Still, 75 of these storms still had enough power to cause injuries and/or damages. The differential impact on the U.S. East Coast clearly reflects the relative position of the East Coast to hurricanes originating in the Atlantic Ocean and traveling west. The number of storm tracks crossing the land area of each state provides a good indication of the relative risk for each state. As might be expected, Florida leads the nation in hurricane experience (defined as the number of times a hurricane or tropical storm has crossed over its state border), followed by North Carolina, South Carolina, Georgia, and Texas (Figure 6-13a). Each of these states has a long history of hurricanes and tropical storms (Barnes 1995, 1998).

The high-risk Atlantic and Gulf coasts are also more prone to loss of life, injuries, and property damage from hurricanes and tropical storms (Figures 6-13b and 6-13c). Florida again tops the list for fatalities, followed by Texas, with 36 and 30 deaths, respectively. According to our data, Florida incurred more than $29 billion in total losses during this time period as a consequence of the tropical storm and hurricane hazard.

A single event has the potential to skew the data and this is clearly reflected in the rankings of hurricane damages. For instance, Hawaii ranks seventh in terms of property damage, even though it ranked much lower for total number of events and deaths. The high rank on damages is a consequence of Hurricane Iniki in 1992, which caused over $1.8 billion in damages. The same is true of other individual hurricane events, notably Andrew (Florida in 1992), Hugo (South Carolina in 1989), and Frederic (Alabama/Mississippi in 1979). Devastating events have not been confined to the southern and southeastern United States. Hurricane Agnes, for example, caused over $2 billion in damage throughout the Northeast in 1972. Note that there is often inconsistency between data sources on the number of fatalities and damages attributed to hurricanes and tropical storms. Again, we must caution that our data represent the most conservative estimates, and thus potentially underestimate the economic losses due to hurricanes and tropical storms.

Earthquakes

The majority of earthquakes (90 percent) occur along the boundaries of tectonic plates (Coch 1995), hence the understood relationship between seismic activity and earthquake hazards in the western United

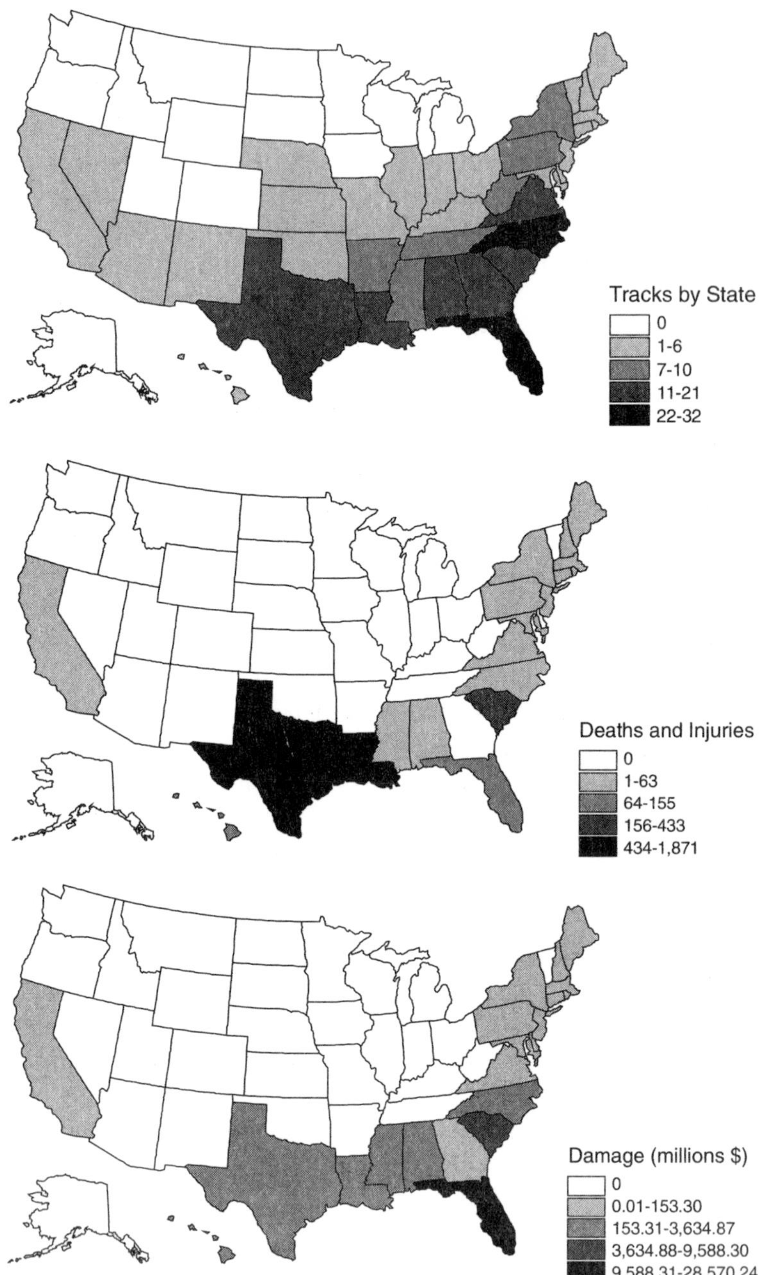

FIGURE 6-13 Trends in hurricane and tropical storm events and losses, 1975-1998: (a) number of hurricane or tropical storm tracks crossing each state, (b) deaths and injuries, and (c) damages (in 1999 dollars).

States. Intraplate seismic activity within the United States is also a concern, however, with historic events occurring near Boston, Charleston (South Carolina), Memphis (Tennessee) and New York City. Whether due to compressional or extensional stress, the resulting crustal break for any of these events releases energy as seismic waves, devastating many building types and infrastructure. Because the quality and quantity of materials and construction varies over time and geographically, so does the amount of damage incurred.

From 1975 through 1998, the Council of National Seismic System's earthquake catalog reveals that the majority of earthquake events (epicenters) occurred in the seismically active areas of the western United States, although there were a number that did affect eastern states (Figure 6-14). The more vulnerable eastern states include those in the vicinity of the New Madrid fault zone (Arkansas, Illinois, Kentucky, Missouri, and Tennessee) and several northern/New England states (New Hampshire, New York, Maine, and Pennsylvania). Tennessee clearly stands out in the Southeast. With nearly 700,000 epicenters, California leads the nation as the most seismically active. Nevada, the next highest, had less than 4 percent of California's total.

Following a geographic pattern similar to that for overall earthquake events, the majority of deaths, injuries, or property damage also occurred in the western United States again with a concentration in California (Appendix B). Several eastern states also experienced significant losses.

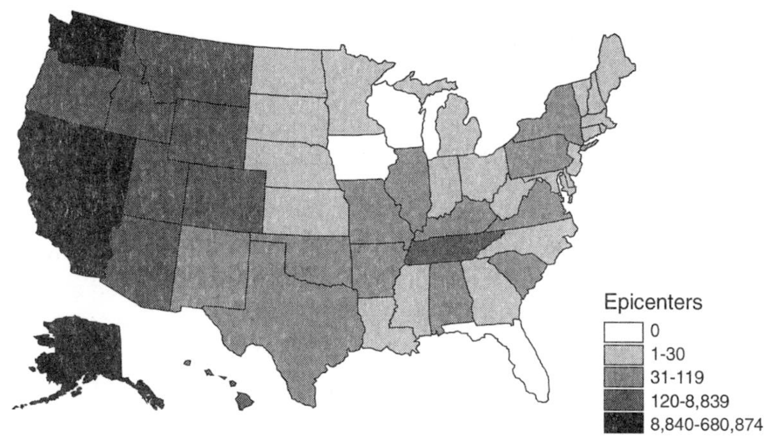

FIGURE 6-14 Earthquake epicenters by state, 1975-1998.

For example, Kentucky had over $1 million (unadjusted) property damage from a 5.1 magnitude (Richter scale) earthquake on July 27, 1980.

Loma Prieta in 1989 and Northridge in 1994 together account for approximately 95 percent of all earthquake damage during the 1975-1998 time period, thus securing California's top ranking for earthquake hazards. Note also that there are serious discrepancies in loss estimates for the Northridge earthquake based on the data source used. Still, losses associated with earthquakes in the United States are not necessarily restricted to "the Big One." A large-magnitude event can occur in a location with few people and thus inflict little loss. A smaller-magnitude event, however, can cause considerable loss if it occurs near a densely populated area. For instance, the Whittier Narrows earthquake was recorded at only a 5.9 Richter magnitude, yet caused over $350 million in damage. Although infrequent, larger events with an epicenter in close proximity to vulnerable populations increase the risk from earthquake hazards and have the potential to produce large damages and significant loss of life.

Volcanoes

Headline-grabbing because of their impressive and explosive releases of energy and heat, volcanoes pose a threat to lives and property from a number of types of hazards—lava flows, lahars, pyroclastic debris, and super-heated and poisonous gases. Earthquakes and tsunamis also may accompany volcanic activity. Nearly all volcanoes occur at the margins of tectonic plates, especially where one plate is subducting another. Other activity occurs within plates where a few "hot spots" (e.g., Hawaii) puncture the plates and send magma to the surface. Forming the eastern edge of the Pacific "Ring of Fire," Washington, Oregon, California, and Alaska are at the greatest risk from volcanic hazards. Other potentially threatened areas in the United States include parts of the Southwest (Arizona, New Mexico) and the Yellowstone area (Montana, Wyoming) (Figure 6-15).

We have very little collective experience with volcanic hazards except in Hawaii. Because the volcanic hazard is so localized and infrequent, there is no national database on deaths, injuries, or damages. Instead, we have case-study data from past events mentioned in the previous chapter (e.g., Mt. St. Helens, Mt. Redoubt, Kilauea). However, as shown in Appendix B, the potential for catastrophic damage and loss of life is significant from this hazard. Since many of the Pacific Coast volca-

FIGURE 6-15 Distribution of volcanoes and volcanic hazards in North America.

noes are located in relatively rural areas and protected as national parks, human and economic losses have been somewhat minimized up to now, especially due to the lack of recent eruptions.

Hazardous Material Spills

Typically, the severity and impact of a hazard event is determined by characteristics such as the event's magnitude or intensity. Further, the nature of the hazard usually is described in terms of loss of life and property, as most of the previous discussion illustrates. The interaction of society, its technology, and natural systems gives rise to technological hazards. Acute events such as nuclear or industrial accidents and oil or hazardous material spills are noteworthy events, as are the more chronic types of technological hazards such as chemical contamination or pollution. Determining the severity of most technological hazards is much more problematic than for many natural hazards, given the unforeseen future consequences (especially human health considerations) that may result from inadvertent exposure. However, we can consider the possibility of exposure as representing a potential risk.

Transportation accidents (highway, rail, air, and water) involving hazardous materials are one of the few technological hazards for which event, death, and damage estimates are collected (Cutter and Ji 1997). Not all incidents result in a hazardous material spill, but the events do represent a likely risk of exposure to toxic chemicals. Although all states have experienced accidents involving hazardous materials, there is a noticeable concentration of events in the eastern half of the United States, especially in the Great Lakes states (Figure 6-16a). Texas and California also ranked among the top five states in terms of accident frequency. The pattern of damages is fairly similar to the distribution of events, except that it is more widespread across the entire United States. Casualties are concentrated in California, Florida, Texas, Montana, and several Great Lakes states (Figure 6-16b). Florida leads the nation in fatalities from hazardous material spills by a factor of two, although, as mentioned in Chapter 5, we believe the Florida source data may have some errors. Damages are concentrated in the eastern half of the nation (Figure 6-16c), although California leads the nation in total dollar damages with $65 million, closely followed by Texas ($63 million). Montana appears in the top five for economic damages because of the 1996 rail spill along Interstate 90 that injured 792 people.

Nuclear Power Plants

Major nuclear accidents in the United States have occurred in both military and commercial sectors. Fermi, Browns Ferry, and Three Mile Island are commercial facilities that experienced accidents during our study period (Cutter 1993). Given the sensitivity of military installations and difficulty in data acquisition, our focus is only on commercial nuclear power plants. Currently, there are 107 operating commercial nuclear power plant units in 32 states (Figure 6-17), which represent 14 percent of the nation's existing electrical generating capacity (EIA 1998). Although the number of nuclear power plants is small in comparison to the other 10,421 U.S. power plants (coal, petroleum, hydroelectric, geothermal, solar, and wind), many suggest that the hazard potential is significantly greater.

There have only been a few nuclear power plant accidents in this country, the most notable being the 1979 Three Mile Island incident (Figure 6-18) (Sills et al. 1982, Goldsteen and Schorr 1991). Since the

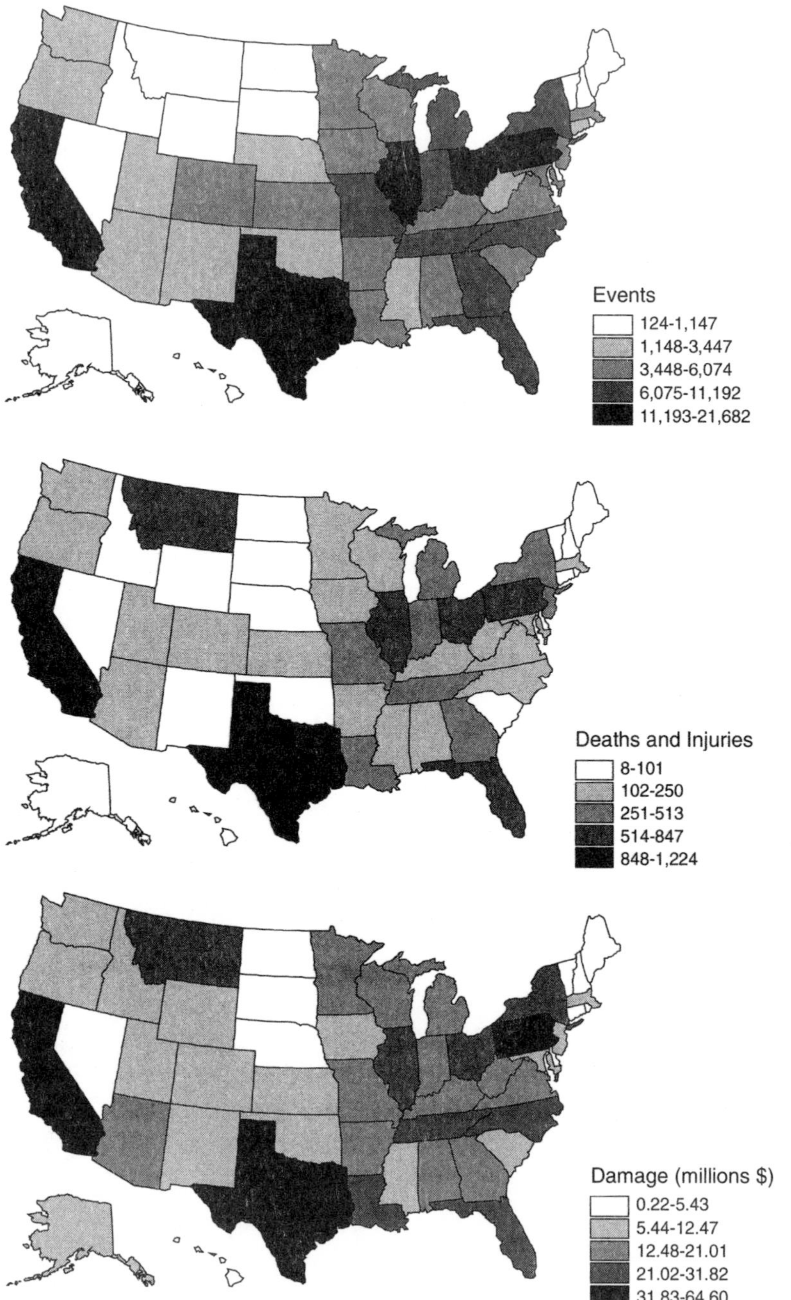

FIGURE 6-16 Hazardous material spills, 1975-1998: (a) total number of spills, (b) deaths and injuries, and (c) damages (in 1999 dollars).

FIGURE 6-17 Commercial nuclear power plants. Note that some facilities have more than one operating unit at the site.

FIGURE 6-18 Three Mile Island, Pennsylvania, where the nation's worst nuclear power plant accident occurred in 1979. Source: Photo by Susan L. Cutter.

number of events is relatively small compared to the operating years of these facilities, a different indicator of the potential hazard is required. The U.S. Nuclear Regulatory Commission (USNRC) provides an assessment of the operation of nuclear power plants in this country in their annual performance review known as the Systematic Assessment of Licensee Performance (SALP) reports (USNRC 1996, 1999). The SALP reports utilized here cover 71 facilities (some facilities have more than one operating reactor) in 32 states between 1988 and 1998. Most of these monitored facilities are located in the northeastern portion of the United States with the exception of Florida and South Carolina. In each report, the facilities were graded and given a score of superior (1), good (2), or acceptable (3) in each of four performance categories (operations, maintenance, engineering, and plant support).

For this discussion, an average score for each category was calculated over the reporting years from the SALP score for each facility. The means were then summed to produce a final score for each state. For example, Alabama (with three facilities) had an average SALP score of 1.28 (Operations), 1.75 (Maintenance), 1.75 (Engineering), and 1.29 (Plant Support) for a final state score of 6.08. These scores do not represent events, but rather the potential for an event based on a performance rating. Higher scores reflect lower safety performance.

There is no real geographic pattern to the rankings (Figure 6-19) although New York and Washington have the worst performance records. Washington has only one operating facility (operated by Siemens

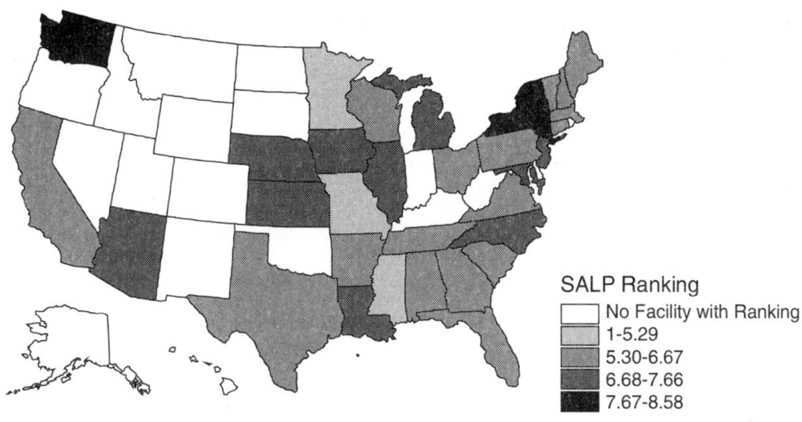

FIGURE 6-19 Average rankings of nuclear power plant performance by state, based on SALP ratings, 1988-1998.

Power Corporation), yet its ranking reveals a relatively poor performance record and thus a greater hazard potential when compared to other states.

Toxic Releases

Representing one of the many potential technological risks that face communities, the Toxic Release Inventory (TRI) provides an indicator of the relative hazardousness of industrial activity. The TRI covers toxic chemicals that are released into the air, water, or land from industrial sources (Lynn and Kartez 1994). The U.S. Environmental Protection Agency (USEPA) only began collecting these data in 1987, and only from selected facilities and covering a limited number of chemicals. Since 1987, the TRI has increased both the types of industries monitored and the chemicals covered. For example, in 1987, around 300 chemicals were covered in the reporting requirements, but by the mid-1990s, more than 600 were routinely monitored. As a result, monitoring changes in the temporal and spatial patterns of both the number of facilities and the total amount of releases is virtually impossible. The releases from TRI facilities depict both direct and hidden costs of pollution and potential health risks from chronic chemical exposures.

Nationwide, there were over 23,000 facilities reporting to the TRI in 1998. These are concentrated in Great Lakes states, Texas, and California (Figure 6-20a). The concentration in the industrial heartland of the nation is no surprise. When the on-site toxic releases are mapped for the year, a very different geographic pattern is apparent (Figure 6-20b). Nevada and Arizona rank highest in toxic emissions (largely due to mining operations). Toxic emissions due to manufacturing are evenly distributed in the eastern half of the nation, posing a significant level of potential risk to the region. Alaska's prominence on the map is due to emissions from paper and pulp mills and electric utilities. Note that the distribution of emissions is based simply on millions of pounds released and does not include a measure of the toxicity of the chemicals being put into the air, water, or onto the land, another indicator of the relative risk posed by this hazard.

Relict Hazardous Waste Sites (Superfund)

Sites contaminated by chemical wastes are the remnants of past dumping practices by companies. Often, the health impacts were unknown at the time and little consideration was given to the long-term

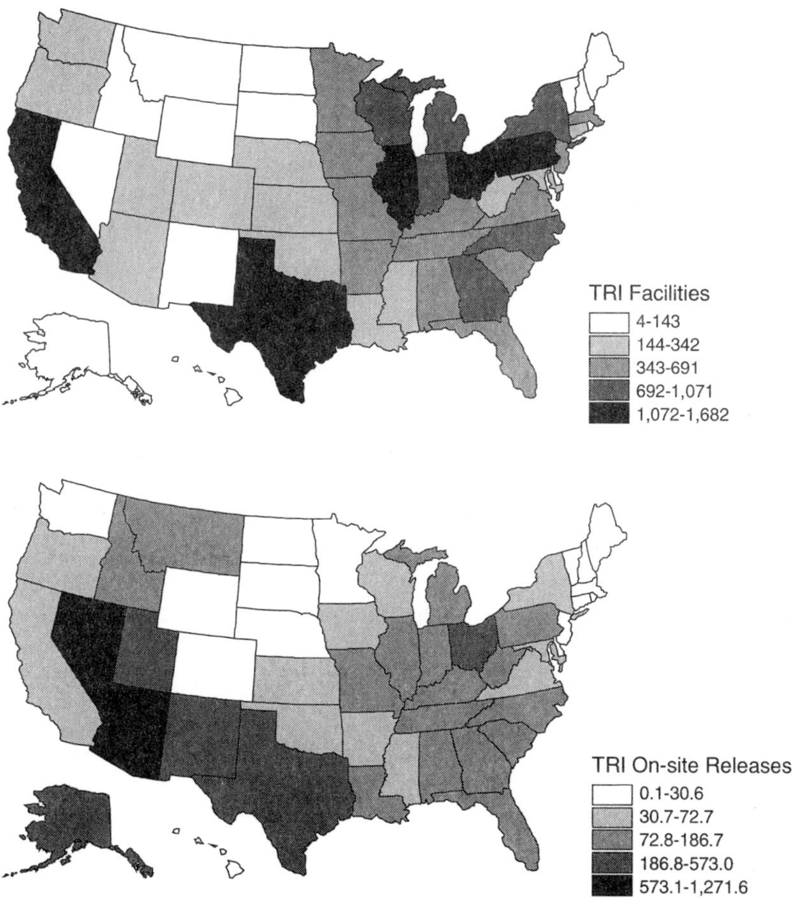

FIGURE 6-20 Toxic Releases, 1998: (a) number of TRI facilities, and (b) releases on site to the air, water, or land in millions of pounds.

environmental consequences. In 1980, Congress passed the Comprehensive Environmental Response, Compensation, and Liability Act (CERCLA, also known as Superfund), which provided the basis for the remediation of these abandoned sites. Not all sites were immediately slated for cleanup, however. Once listed as a potential source of contamination, the site goes through a listing and evaluation process. If the contamination is severe enough and poses imminent danger to human health, it is placed on the National Priority List (NPL) for immediate remediation. The NPL thus represents those sites that have the greatest potential to harm human health and the environment. Part of the

Superfund legislation also authorizes the search for responsible parties in order to recover some of the state and federal costs of cleanup. However, the costs associated with remediation of these abandoned waste sites since 1980 is well into the billions of dollars (Hird 1994).

Not all states bear the same environmental burden when it comes to hosting NPL sites. Of the 1,245 sites currently listed, New Jersey has the most with 113 and North Dakota the least with none. Similar to the patterns observed with other technological hazards, the concentration of NPL sites is in the Northeast, especially New York, Pennsylvania, and New Jersey (Figure 6-21). Florida, Michigan, Texas, California, and Washington also host a substantial number of NPL sites.

REGIONAL ECOLOGY OF DAMAGING EVENTS

This chapter began with a series of maps illustrating the geographic distribution of overall losses for the nation. On one level, these maps reveal which states have the greatest loss of life and property and thus which ones may require some type of intervention to reduce the future impacts of hazards. One of the ways in which the federal government assists counties and states in a time of crisis is to issue disaster declarations under the Stafford Act. In fact, when the number of disaster declarations is mapped by individual state for the 24-year period, the geo-

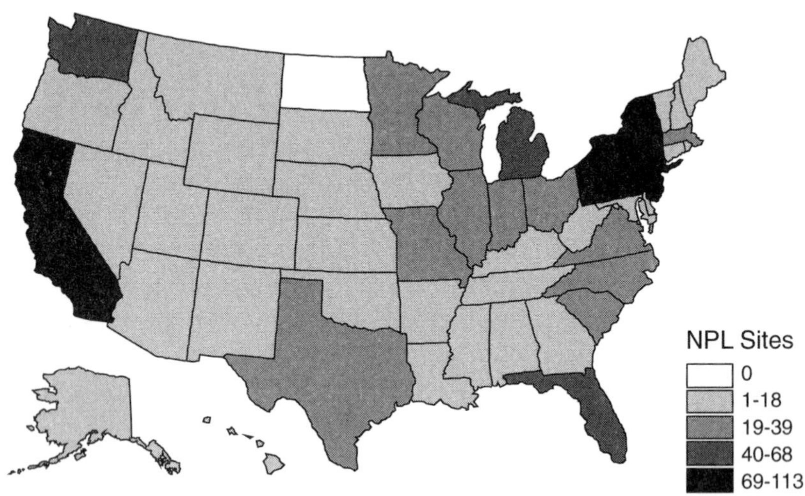

FIGURE 6-21 Number of Superfund NPL sites, 1998.

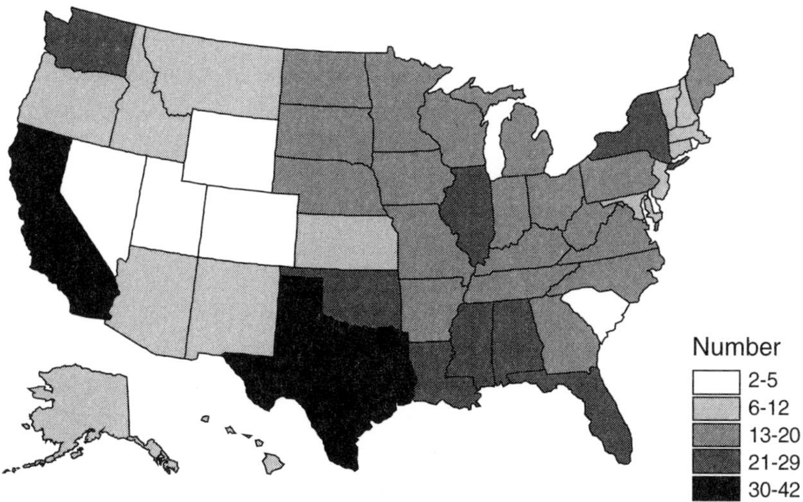

FIGURE 6-22 Presidential disaster declarations, 1975-1998.

graphic pattern (Figure 6-22) shows some similarities (Gulf Coast states and California) but some interesting differences. Note, for example, that South Carolina had comparatively large total damages during this period (Figure 6-1b), yet received very few disaster declarations (4). Conversely, Washington received a large number of disaster declarations (26) (Figure 6-22), but our damage estimates for the state placed it in the lowest category for loss (Figure 6-1b). California and Texas had the highest number of presidential declarations (42 and 45, respectively), and they also had the highest total dollar amounts in damages. Those states with the least property and crop loss damages due to natural hazards according to our data also had the lowest number of disaster declarations. The Southeast (except South Carolina) and the Gulf Coast states, especially Florida, had a relatively high number of declarations as did New York, Illinois, and Oklahoma.

Although examining the geographic distribution of raw damages and the total number of deaths has some utility, it does not uncover the differential burdens of a state's disaster experiences. To begin to understand the complexity of loss distribution in the United States, we normalized damage losses by number of people in the state, by area of the state, and by the state Gross Domestic Product (GDP) in 1997. The number of deaths attributed to disasters was also recalculated as a rate per 100,000 people to account for the effect of large populations.

Economic Losses

Loss per capita shows the relative economic burden for each individual residing in the state. In theory, huge losses in heavily populated states become more widely distributed and thus compare favorably to fewer losses in less populated areas. The map of per capita hazard losses (Figure 6-23a) shows a very different pattern from the raw loss totals (Figure 6-1b). California and Texas (heavily populated, with large losses) no longer rank at the top. Instead, North Dakota now has that distinction. Regionally, states in the Mississippi River Valley and Gulf Coast rank quite high. Even with a fairly large population, Florida still ranks ninth; South Carolina and Idaho (with lower populations) also are among the top ten states for per capita damages (Table 6-3).

Another way to examine losses is through some measure of areal density, which takes into account the sheer physical size of the state (Figure 6-23b). The density control enables us to more effectively compare very large states (Texas and California) with very small ones (New Jersey and Connecticut). In this configuration, Florida, Hawaii, and South Carolina get top honors as the most disaster-prone states, with losses exceeding $400,000 per square mile during our study period (Table 6-4). Alaska and the Rocky Mountain states have the lowest damage density—not a surprise given the large size of those states.

Finally, standardizing economic losses by state GDP (1997) provides an indication of a state's ability to financially recover from disasters (Figure 6-23c). In fact, this map is considerably different not only from the maps of raw damage, but from the per capita map as well. Losses standardized by state GDP also catapults North Dakota to the position of the state most impacted from hazards (Table 6-5). California and Texas slide further down in the rankings due to their large wealth and thus ability to absorb the losses and recover. Poorer states such as South Carolina, Mississippi, Idaho, and Iowa appear to be the most disaster-prone places on this indicator, based on the potential economic impact of all hazards on these states' economies.

Pattern of Disaster Deaths

Using another indicator of loss—death—the most hazardous places do not necessarily coincide with those states with the most economic damage. In fact, a very different geographic pattern emerges, as shown previously (Figure 6-1a). However, if you normalize the total number of

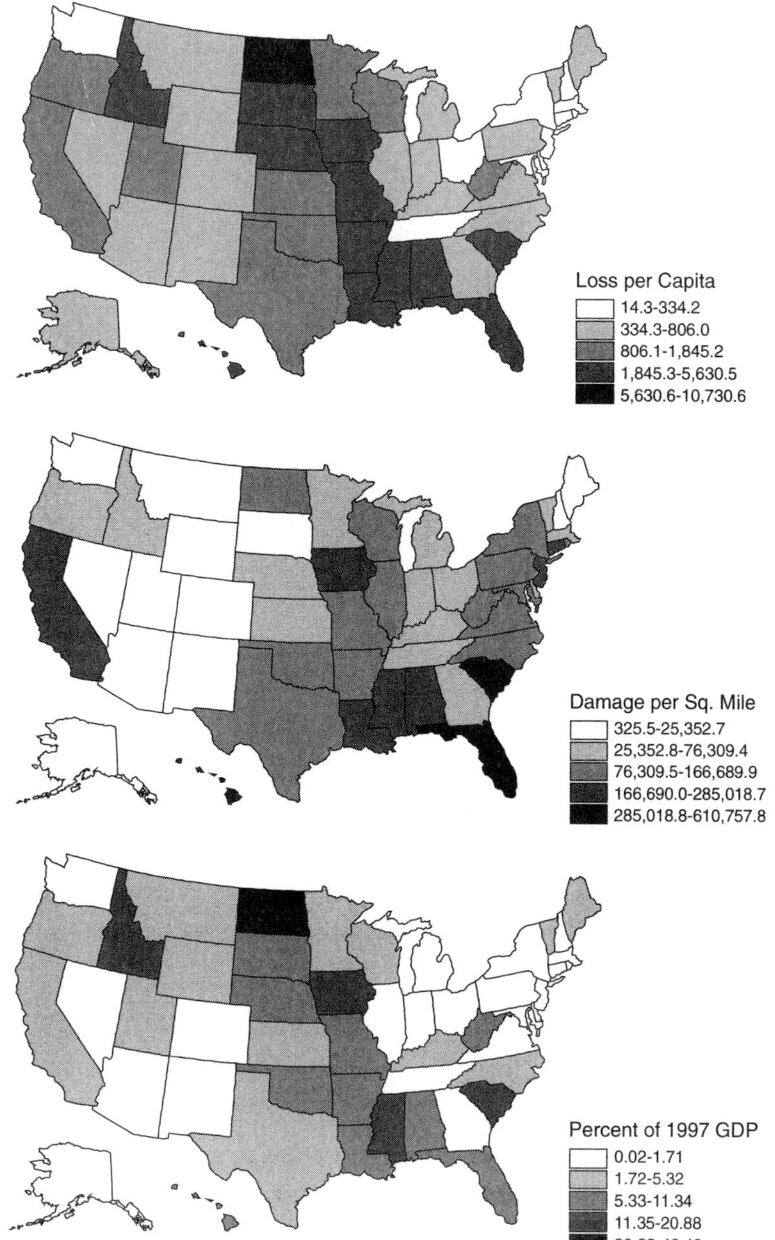

FIGURE 6-23 Normalized total economic damages from all hazards, 1975-1998, based on (a) dollar losses per capita, (b) dollar losses per square mile, and (c) losses as a proportion of 1997 state GDP.

TABLE 6-3 Disaster Losses per Capita, 1975-1998

State	Dollars per Person
North Dakota	10,631.53
Iowa	5,610.06
Mississippi	4,727.64
Idaho	3,704.44
South Carolina	3,459.23
Louisiana	2,939.23
Alabama	2,891.75
South Dakota	2,732.97
Florida	2,548.46
Hawaii	2,493.54
National average	1,214.98

TABLE 6-4 Top 10 States for Damage, by Density, 1975-1998

State	Dollar Amount of Damages per Square Mile
Florida	610,758
Hawaii	430,237
South Carolina	400,755
Louisiana	285,019
Iowa	279,811
Mississippi	259,535
California	257,669
Alabama	230,361
Connecticut	207,680
New Jersey	200,294

TABLE 6-5 Property and Crop Losses, as a Proportion of the 1997 State GDP, 1975-1998

State	Losses as a Percentage of State GDP
North Dakota	43.42
Mississippi	20.88
Iowa	19.43
South Carolina	12.94
Idaho	12.92
Alabama	11.34
Louisiana	9.99
Arkansas	9.85
South Dakota	9.53
Florida	8.66
National average by state	5.08

deaths attributed to all hazards as a rate (in other words number of deaths per 100,000 people) rather than just raw numbers, a very different regional pattern emerges (Figure 6-24). The national fatality rate is 4.4 deaths per 100,000 people. Four states have a high death rate (more than two to three times the national average) from all hazards—Alabama, Arkansas, Colorado, and Wyoming (Table 6-6). Only two of these (Colorado and Alabama) were in the top ten for total number of deaths. With the normalization procedure, we can moderate the influence of population size and determine the relative risk of death for each state. The low fatality rate in California is interesting given the high economic

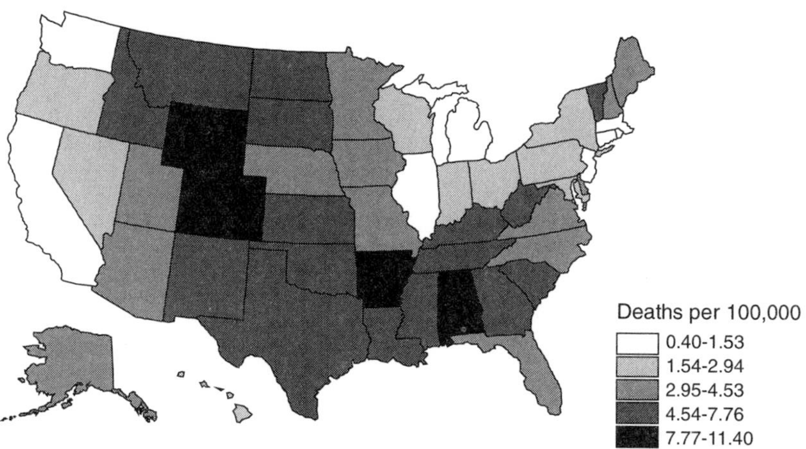

FIGURE 6-24 Hazard deaths per 100,000 people, 1975-1998.

TABLE 6-6 Deaths per 100,000 Attributed to Hazards, 1975-1998

State	Death Rate (deaths per 100,000)
Arkansas	11.40
Alabama	10.89
Colorado	8.95
Wyoming	8.82
Montana	7.76
Tennessee	7.52
North Dakota	7.36
Mississippi	7.03
West Virginia	7.03
New Mexico	7.00
National average	4.41

damages from hazards. Also, Texas (labeled "the fatality state" earlier on the basis of raw data) drops to eighteenth when population size is considered.

Hazards of Places

As was done in *Disasters by Design* (Mileti 1999), we calculated a simple composite measure of the relative hazardousness of states, using three indicators—number of hazard events, casualties (deaths and injuries), and losses. The event score reflects the potential risk from hazards for which we have no death or damage estimates, such as toxic releases. The level of risk from other hazards is captured by our casualty and damage estimates. Each indicator was constructed by assigning each state a value based on the percentage of that state's contribution to the national total, be it events, casualties, or dollar losses. The percentages were then summed for all hazards for each state to derive an average, which was then ranked and mapped. In this way, we can compare and geographically examine the relative hazardousness of individual states and/or regions. As can be seen in Figure 6-25 the most disaster-prone states are Florida, Texas, and California, where number of events, damages, and casualties are extremely high. Florida and California rank high because of a number of large catastrophic events, whereas the pattern for

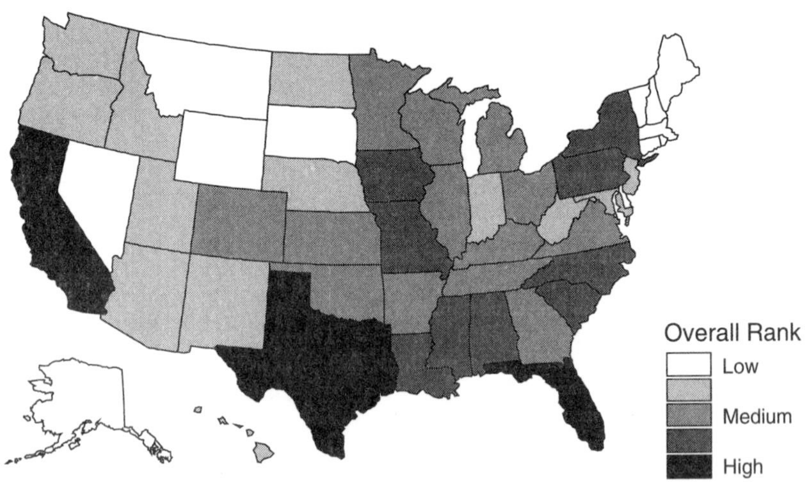

FIGURE 6-25 Hazard proneness by state, based on risk potential and overall losses.

Texas is reflective of the sheer number and diversity of hazardous events during the study period. Note the large number of states in the low category—states that have quite a bit of exposure to hazards, but where the losses (economic and casualties) do not reflect more catastrophic events (Table 6-7).

CONCLUSION

The American hazardscape is a multifaceted quilt of hazard events (and thus exposure), casualties, and economic losses. The common perception of which states are the most hazardous may not always fit the reality of the data. The level of hazardousness and the spatial variability depend on the type of hazard and specific indicators of losses (casualties and economic damages). For example, there is an inconsistent pattern

TABLE 6-7 Overall Hazard Scores

State	Event	Death	Damage	Overall	State	Event	Death	Damage	Overall
FL	3.25	47.34	113.18	54.59	IN	2.21	13.44	7.02	7.56
TX	3.96	88.75	61.49	51.40	AZ	4.73	13.68	3.80	7.40
CA	4.46	39.08	102.36	48.63	ND	0.09	4.30	17.44	7.28
AL	2.05	40.29	32.56	24.96	WV	0.71	11.33	5.91	5.98
SC	2.83	22.59	40.56	21.99	OR	0.68	6.98	10.21	5.95
LA	2.04	23.72	33.24	19.67	NE	1.16	6.02	10.27	5.82
NY	4.56	37.87	12.20	18.21	MD	1.08	11.33	3.05	5.16
MO	2.00	19.52	29.35	16.96	ID	0.57	5.18	9.59	5.11
IA	1.32	9.31	39.89	16.84	NJ	4.14	6.21	3.81	4.72
MS	1.00	15.76	33.73	16.83	UT	2.54	6.79	4.80	4.71
PA	5.80	31.29	12.68	16.59	HI	0.13	3.09	9.30	4.17
NC	2.65	29.06	14.08	15.27	NM	1.22	9.07	1.62	3.97
GA	2.24	29.90	7.34	13.16	WA	1.72	4.85	4.06	3.55
AR	1.16	23.44	14.80	13.13	NV	4.77	2.42	2.03	3.07
TN	2.20	32.78	4.20	13.06	SD	0.18	3.81	4.92	2.97
IL	5.25	15.33	17.16	12.58	MT	0.70	5.51	1.56	2.59
OH	4.22	25.32	8.04	12.53	CT	2.07	2.49	2.66	2.41
OK	0.72	18.11	14.88	11.23	ME	0.73	4.79	1.61	2.38
CO	0.62	25.99	4.42	10.34	MA	1.55	3.67	1.64	2.29
VA	2.10	17.93	8.93	9.65	NH	0.87	3.70	0.43	1.67
WI	2.64	12.55	13.06	9.42	VT	0.53	3.40	0.80	1.58
KY	1.18	17.36	7.13	8.56	WY	0.16	3.53	0.94	1.54
MI	4.20	11.90	8.73	8.28	AK	1.29	1.50	0.47	1.09
MN	1.88	11.67	11.23	8.26	DE	0.48	1.78	0.26	0.84
KS	0.98	13.88	9.79	8.22	RI	0.41	0.65	0.36	0.47

between actual damages experienced by states (1975-1998) and the number of presidential disaster declarations during this same time period. Given the politicized nature of presidential declarations and the need to meet minimal threshold requirements on a disaster-by-disaster basis, this is not surprising. The American hazardscape consists of large singular events for some states (Florida and California), but more often than not, the regional pattern of hazard losses for the nation is composed of frequent, less costly events, that, over time, add up to billions of dollars in direct economic losses and untold numbers of injuries and deaths. You stand a greater chance of dying from a natural hazard event if you live in Arkansas, Alabama, Colorado, or Wyoming than if you live in California, Texas, or Florida. The per capita losses during this period averaged around $1,200 for every person in the nation. North Dakotans experienced per capita losses nine times the national average, the highest in the country. While there is significant federal interest and funding in earthquake research, planning, and mitigation, weather-related disasters are more costly to this nation than seismic ones, with drought posing the most significant problems for the future. However, the catastrophic potential for earthquake losses throughout the nation remains very high.

Since we are a nation obsessed with rankings, we include our contributions to this American pastime, based on our 24-year study period. The most exposed state (based on potential risks from hazards, not actual losses) is Pennsylvania (because of the large amounts of toxic releases, poor nuclear power plant performance, and relict hazardous waste sites). Texas has the largest number of fatalities, yet when computed as a rate (to account for large populations), Arkansas lays claim to the title of "hazard fatality state." California has experienced the most economic losses (based on 1999 dollars). Yet when we consider both population and the size of the state, California's supremacy declines. Instead, Florida ranks number 1 for greatest economic losses ($611,000) per square mile, whereas North Dakota claims that honor based on per capita economic losses ($10,600 per person).

Perhaps the most telling statistic is the impact of hazards on a state's economic well-being (measured as a percentage of its GDP). The state most economically affected by hazards is North Dakota, where total losses during the past 24 years amounted to 43% of its 1997 state GDP.

When all indicators (events, casualties, economic losses) are combined, Florida, Texas, and California are the most hazard-prone states, whereas Rhode Island, Delaware, and Alaska are the least hazard-prone. It is the combination of multiple types of hazards, people living in hazardous locations (coastal areas, floodplains, seismic zones), along with large urban populations that contribute to economic losses and casualties from hazards and thus define the nation's hazardscape. We must learn the lessons from these historic and geographic trends in hazard events and losses and use this knowledge in developing more sustainable options for vulnerability reduction and hazards mitigation throughout the nation.

CHAPTER 7

Charting a Course for the Next Two Decades

Susan L. Cutter

The American hazardscape stretches from coast to coast and from border to border. There are very few states that are not affected in one way or another by environmental hazards—be it a flood, severe weather, seismic activity, or a hazardous material spill. There is no such thing as a "hazard- or disaster-free" environment, but there are clearly places that are more hazard-prone than others. What accounts for the geographic distribution of hazard events and losses and how can we reduce the impact of hazards and disasters in the future? What actions must local communities and the nation take to reduce our current and future vulnerability to environmental hazards?

TAKING STOCK

The conservative estimate of losses used in this book presents a grim picture for the nation during the past two and a half decades. Dollar losses have been escalating, especially during the 1990s. During the past 24 years, economic losses averaged more than $12 billion annually. The 1990s were the costliest decade on record ($17.2 billion and still counting), and 1992, 1993, and 1994 the costliest years ever. Economic losses in the mid-1990s were almost double those of a decade earlier. To place this in perspective, these cumulative losses represent about 4 percent of the nation's Gross Domestic Product in 1997. We can say without equivocation that hazard losses are escalating.

Less conservative estimates, for example, suggest that the figures range from $54 billion annually in direct losses (during the past 5 years) to over $160 billion (in direct and indirect damages) annually between 1988 and 1998. Although we can argue about the magnitude of the losses and what they cover (direct, indirect, or hidden costs), the trend remains clear—expect greater economic losses from environmental hazards in the future.

Although the trend in human injuries and fatalities has improved somewhat over time, more than 350 people die every year as a direct result of environmental hazards. Floods, lightning, and tornadoes caused the most fatalities, historically, but on an annual basis, heat waves result in the greatest loss of life by an individual hazard. It is not the singular catastrophic event (such as Hurricane Andrew or the 1993 Midwest floods) that is solely responsible for the current snapshot of the American hazardscape. Rather, it is the cumulative impact of frequent, yet lower consequence events (a winter snowstorm, a midsummer heat wave) that contribute to an overall national pattern of hazard events and losses. This is not to downgrade our concern with the potentially catastrophic event. Instead, it highlights the need for increased monitoring, preparedness, and mitigation of these more pervasive and less catastrophic events.

As suggested by others (Changnon and Easterling 2000, Easterling et al. 2000) an increase in atmospheric extremes is partially contributing to increased losses during the past two decades. Equally important, however, if not more so, is the increasing movement of people, infrastructure, and investments into hazardous areas—such as coastal areas or seismically active regions. Our vulnerability has increased during the past two decades and will continue to do so, with or without an increased frequency of hazard events.

So what can be done to reduce the impact of hazards and disasters on society? First and foremost, a shift in public policy is required. We need to move away from a disaster assistance mindset (rewarding individuals and communities for building and living in hazard-prone areas) to one that fosters long-term thinking about mitigation, loss reduction, and personal responsibility for actions. As suggested in the findings of the second assessment of hazards (Burby 1998, Kunreuther and Roth 1998, Mileti 1999) and others (Platt 1999), we need to work toward sustainable and disaster-resistant communities. Specifically, we need to defederalize the costs of disasters and limit federal subsidies on risk. We need to make sure that individuals and communities who choose to build in known hazardous environments take responsibility for the risk they assume. Second, we need to improve hazards mitigation and reduce our vulnerability through wiser land-use decisions and ensure that land-use policies are integrated so as to preclude development and redevelopment of hazardous environments.

Achieving sustainable and disaster-resistant communities will take time. A first step toward that goal, based on a very pragmatic concern, is to establish our current level of hazards vulnerability, describe how this has differed from the past level and what we might anticipate in the future. To accomplish this, we need to immediately implement two of the recommendations of the Second Assessment (Mileti 1999):

- Conduct a nationwide hazard and risk assessment:
  Not enough is known about the changes in or interactions among the physical, social, and constructed systems that are reshaping the nation's hazardous future. A national risk assessment should meld information from those three systems so hazards can be estimated interactively and comprehensively, to support local efforts on sustainable mitigation (p. 11).
- Build national databases:
  The nation must collect, analyze, and store standardized data on losses from past and current disasters, thereby establishing a baseline for comparision with future losses (p. 12).

VULNERABILITY SCIENCE

We have not adequately developed the integration between natural sciences, engineering sciences, and social sciences to produce credible vulnerability assessments at the local level. Vulnerability science is still in its infancy, yet there are major parallels with the development of hazards research almost 25 years earlier. Both fields bring together researchers (from many disciplines) and practitioners who are interested in reducing hazard losses, who recognize the complex nature of the task, and who understand the need for different approaches, methods, and tools in effecting policy changes.

There is considerable research activity addressing various components of vulnerability (social systems, estimations of risk, infrastructure), but there is no standard technique for determining vulnerability from a range of hazards, as we demonstrate in Chapter 2. There are specific hazards assessments, ranging from locally based studies to global perspectives, from single-hazard sources to multiple hazards. There is only peripheral interest in differential vulnerabilities, that is, the potential impact of hazards and disasters based on gender, wealth, race/ethnicity, and age. In fact, rarely are data systematically collected on the impacts of disasters on various subpopulations.

As a tool for representing vulnerability, many studies employ some type of map based on a geographic information system to display the results. Advances in mapping and in the geographic information sciences have improved our understanding of risk and vulnerability. The Internet has helped to widely disseminate hazards information. Yet, we are still plagued by both conceptual and pragmatic issues. First, we lack consistent and comparable information on exposure or risk indicators. Our science is not at the stage where we can model, let alone predict, future exposures with any degree of certainty. Second, we need a similar set of social indicators that adequately describe the social vulnerability of people and the built environment at the local level. Finally, we need improvements in our capability to analyze and visualize these results.

Federal research programs emphasize seismic risks, yet greater losses result from hurricanes, severe weather, flooding, and drought. Population vulnerability is potentially greater in regions affected by these hazards as well. We need to build our research capacity in the area of vulnerability science. This necessitates support not only for the natural science community, but engineering and social sciences as well. We need to develop the scientific infrastructure (models, methods, and tools) to integrate environmental exposure/risk with the social vulnerability of local communities. In this way, we will be able to produce hazards assessments that can be incorporated into the local planning processes (master plan) throughout the nation and thus inform local decision making about land use.

NATIONAL DATABASE ON HAZARD EVENTS AND LOSSES

We have fragmented and incomplete data on hazard events and losses. Without this basic information, we cannot even begin to improve hazards and vulnerability assessments. Hazard events data are collected by many different state and federal agencies (NOAA, USGS, FEMA, EPA), researchers, and the private sector (primarily insurance companies). The federal mission agencies collecting hazards-related data have a broad range of mandates, which are reflected in the type of data they compile. For example, during a major hazard event, each federal agency has responsibility for a segment of the response, impacts, or losses, as specified under the Federal Response Plan. Unfortunately, archiving much of the data is not mandatory, is not funded, and, as such, valuable information is typically lost in the months or years after the event.

Some of the agencies are more interested in specific types of event information (earthquakes, hurricanes), but do not always collect data on losses. Others are more concerned with loss information (property damage, insurance claim payouts) and do not focus on specific hazard events or risk. The time frame for the existing data sets (as pointed out earlier in this volume) varies substantially, limiting historic comparisons. Lastly, geographic resolution is an issue for many of the data sets currently used.

Data sharing or collaboration is an infrequent event, and so, when a hazard falls in the domain of two federal agencies (or within different branches of the same agency), the result is a fragmented and often incomplete database. Historic trends in hazard events are not systematically archived because it is often beyond the agency's mandate. We have become so overly specialized both in our agencies and in academe that trying to pull together an "all-hazards" approach or team is almost impossible. However, there are some good event databases, which we describe in Chapter 4, most notably those on seismic and atmospheric events.

The problems with loss data sets are more critical. The definitions of "loss" are often inadequate, and we have difficulty in standardizing what is meant by loss. As noted by Gilbert White, "Accurate and comparable data on losses of lives and property from extreme events are very difficult to assemble; standards and methods of data collection are far from uniform and consistent" (1994:1237).

Mission agencies, although able to monitor hazard events, often lack the infrastructure and capability to provide damage and/or loss estimates. Direct versus indirect, public uninsured versus private insured—all need to be included when we describe what hazards and disasters cost this nation. Currently, they are not. Loss estimation is often more of an "art" than a science. In response to the preliminary findings of Mileti (1999), the NRC (1999b) proposed an initial framework for systematic data collection for determining hazard losses. The Heinz Center (2000a) also provided a foundation for examining direct, indirect, and hidden costs associated with one hazard—coastal erosion—which could be seen as a prototype for a national loss inventory.

We are the most advanced and wealthiest nation in the world and to find the nation does not have a systematic accounting of hazard losses by specific hazard source by location for the entire country is an appalling situation. Our conclusion is simple. There is no national commitment to loss reduction assessments; if there were, these data would be readily available so that we could monitor our progress toward that goal. Although the task may seem daunting, it does not obviate the need for this national data set. Rather, it makes the compilation of such a national inventory an essential and important task.

NATIONAL LOSS INVENTORY/NATURAL HAZARD EVENTS DATA CLEARINGHOUSE

Once we have this inventory of hazard events and losses, we must make the information available to researchers and practitioners. Therefore, we need to establish a National Loss Inventory/Natural Hazard Events Data Clearinghouse. The goals of the clearinghouse are to (1) facilitate the creation of the national hazard event and loss inventory and (2) serve as the data archive or repository for current and historical hazard events and loss data. The clearinghouse would assist in research efforts to develop better loss estimation metrics and better exposure data. Where data are currently unavailable, the clearinghouse should have the authority to work with agencies in producing the necessary data collection tool and then collect the data. It would monitor trends over time and geographic location and thus provide the initial infrastructure for baseline vulnerability assessments at the local level. The clearinghouse would be the central repository for event and loss data with storage, catalog, geocoding, and metadata functions—all the tools and procedures necessary to respond to the most basic inquiries regarding hazard losses.

The National Loss Inventory/Natural Hazard Events Clearinghouse must take advantage of current technology to provide on-line (or Web) access to information by researchers, practitioners, decision makers, and the general public. There should be a query capability so that users can search by topic or geographic location ranging from census blocks to congressional districts, to entire states. All data should adhere to Federal Geographic Data Committee (FGDC) standards and include the requisite documentation for any spatial database.

Finally, the clearinghouse should not be housed within a single mission agency, but rather operate independently and draw on partnerships with primary data providers (federal agencies and the private sector). One model would be to expand the purview of the current Council on Environmental Quality (CEQ) and rename it the Council on Environmental Quality and Hazard Reduction to include the compilation and archiving of hazard event and loss data. Another model would be to task an executive branch agency with these responsibilities. For example, the Federal Emergency Management Agency (FEMA) currently has an extremely limited "in-house" research or planning capability. The establishment of an expanded office (similar to what the U.S. Environmental Protection Agency [USEPA] has done) would help FEMA's longer-term mission as it moves its orientation from a purely response mode (after the disaster) to a mode emphasizing preparedness, mitigation, and most importantly, vulnerability reduction. Irrespective of its administrative location, the clearinghouse should issue an annual "The State of Disaster" report (analogous to CEQ's *State of the Environment Annual Report*) highlighting trends in hazard events and losses, specific improvements (or lack thereof) in loss reduction, and looming issues, thus providing the blueprint and scientific basis for future policy changes.

THE WHOLE IS GREATER THAN THE SUM OF THE PARTS

Like the summary volume (Mileti 1999), this book is more than just a call for an improved hazard event and loss database and a mechanism to disseminate the data. It is a call for a new way of thinking about how we monitor, assess, and ultimately reduce our vulnerability to environmental threats. The path we need to take is obvious, and unless we change course now, we can expect that losses from environmental hazards will increase. The only issues for debate are (1) by how much and (2) where?

The foundations for our future path consist of five different dimensions. First, we must provide a consistent and integrated set of data (and information) in support of public policies. We need to integrate social systems, environmental systems, and the built environmental systems in ways that have not been done before. We need to ensure that the differing agencies collecting data based on each and all of these systems define losses in the same way and calculate damages consistently. We need to take the long view in our understanding of hazard events and losses and examine not only past and present trends and opportunities, but future ones as well. We need to be proactive and forward thinking about our data needs and anticipate future surprises, whatever they may be.

Second, we need to consider distributive justice in our public policies aimed at vulnerability reduction. There were, are, and will be inequities in the patterns of losses based on geographic location and demographic characteristics, and these inequities will be faced by the next generation as well unless we implement policies and programs to redress the present situation. This may require substantial resources in some places or for some socioeconomic groups in order to reduce vulnerability and enable the development of more sustainable communities. How we address the issue of intragenerational equity in hazards reduction will be an important public policy question in the next decade.

Third, we need a strategic plan for hazards reduction at all levels of government. This plan should be based on tangible goals and specific indicators of accountability. Do expenditures for hazards reduction programs make a difference, or do they facilitate the movement of people into increasingly hazardous areas? We need audits of our national disaster aid, recovery, and insurance programs to assess their effectiveness in reducing losses and overall vulnerability. Along the same lines, we should have an in-depth analysis and report for each presidential disaster declaration, providing very detailed information on losses (current and future), opportunities for mitigation, and lessons learned. We also need to increase the qualifying threshold for receiving presidential disaster declarations in order to de-politicize the process and make the declarations more consistent with large losses.

Fourth, we need to support research and public outreach programs on a wide range of environmental hazards, not just the natural hazard "du jour." The seismic hazard research program is funded fairly well by the National Science Foundation, for example, yet there is not the same level of commitment to weather-related hazards and their impact on society. We need to develop the necessary conceptual frameworks, models, data, and tools to assist in policy choices about hazards reduction strategies in the future. We need to assess the relative impact of hazard events and losses geographically and socially to see if there are certain regions or segments of society that are disproportionately affected, thus reducing their collective resilience to future disaster events. We cannot do that without advancements in vulnerability science. As we begin to understand more fully the cause and consequences of losses and where they occur, resources for research and outreach should be prioritized and allocated accordingly.

Finally, we need to develop a collective national consciousness that places the right to a sustainable quality of life on par with economic security and emotional well-being. Reducing the nation's vulnerability to environmental hazards will take public support and political will and needs to be addressed within the confines of local community growth and economic development constraints and national priorities for hazard mitigation. The federal government should not bail out communities and individuals when they make foolish locational decisions. Instead, the burdens will be placed locally, and so will the solutions.

Improvements in vulnerability science and in reducing losses from disasters will rest on the compilation of adequate data to monitor our successes and our failures and the political will to make tough choices. Without a national inventory on hazard events and losses, we have no baseline on current resources spent (or in some cases respent) to keep people out of harm's way, nor do we have any rational basis for planning future "disaster-resilient" communities. Reducing our vulnerability to hazards should be based on public policies guided by the best science, information, and data available at the time, not political expediency. To do otherwise limits our ability and that of the next generation to lessen the impacts of environmental hazards on people and places.

References

Abrams, M., E. Abbott, and A. Kahle. 1991. Combined use of visible, infrared, and thermal infrared images for mapping Hawaiian lava flows. *Journal of Geophysical Research* 96:475-484.

Algermissen, S. T., and D. M. Perkins. 1976. A Probabilistic Estimate of Maximum Acceleration in Rock in the Contiguous United States. USGS Open File Report 76-416. Reston, VA: USGS.

Algermissen, S. T., D. M. Perkins, P. C. Thenhaus, S. L. Hanson, and B. L. Bender. 1982. Probabilistic Estimates of Maximum Acceleration and Velocity in Rock in the Contiguous United States. USGS Open File Report 82-1033. Reston, VA: USGS.

Aller, D. 1999. A reinsurance perspective of risk assessment. Pp. 252-253 in *Natural Disaster Management*, J. Ingleton, ed. Leicester, England: Tudor Rose.

American Institute for Economic Research. 1999. *Economic Education Bulletin* Vol. XXXIX (12):4.

Anonymous. 1997. Tornado forecasting and warning. *Bulletin of the American Meteorological Society* 78:2659-2662.

Anonymous. 2000. FEMA Says Floyd Claims Second Highest. *Natural Hazards Observer* XXIV (5):13-14.

AWMA (Air and Waste Management Association). 1997. Are inaccurate emission measurements clouding pollution control planning? *EM* (September):16-17.

Barnes, J. 1995. *North Carolina's Hurricane History*. Chapel Hill, NC: University of North Carolina Press.

Barnes, J. 1998. *Florida's Hurricane History*. Chapel Hill, NC: University of North Carolina Press.

Barrows, H. H. 1923. Geography as human ecology. *Annals of the Association of American Geographers* 12:1-14.

Barry, J. M. 1997. *Rising Tide: The Great Mississippi Flood of 1927 and How It Changed America*. New York: Simon & Schuster.

Beardsley, T. 2000. Dissecting a hurricane. *Scientific American* 282(3)(July):80-85.

Been, V. 1995. Analyzing evidence of environmental justice. *Journal of Land Use and Environmental Law* 11(1):1-36.

Beroggi, G. E. G., and W. A. Wallace, eds. 1995. *Computer Supported Risk Management*. Dordrecht, The Netherlands: Kluwer Academic.

Berry, B. J. L. 1977. *The Social Burdens of Environmental Pollution*. Cambridge, MA: Ballinger Press.

Blaikie, P., T. Cannon, I. Davis, and B. Wisner. 1994. *At Risk: Natural Hazards, People's Vulnerability, and Disasters*. London: Routledge.

Bomar, G. W. 1995. *Texas Weather*. Austin: University of Texas Press.

BOND (Board on Natural Disasters). 1999. Mitigation emerges as major strategy for reducing losses caused by natural disasters. *Science* 284:1943-1947.

Brainard, J. S., A. Lovette, and J. P. Parfitt. 1996. Assessing hazardous waste transport risks using a GIS. *International Journal of Geographical Information Systems* 10:831-849.

Bryant, E. 1991. *Natural Hazards*. New York: Cambridge University Press.

Burby, R. J., ed. 1998. *Cooperating with Nature: Confronting Natural Hazards with Land-Use Planning for Sustainable Communities*. Washington DC: Joseph Henry Press.

Burton, I., and R. W. Kates, eds. 1986. *Geography, Resources and Environment: Vol. I, Selected Writings of Gilbert F. White; Vol. II, Themes from the Work of Gilbert F. White*. Chicago, IL: University of Chicago Press.

Burton, I., R. W. Kates, and G. F. White. 1993. *The Environment as Hazard* (2nd ed.). New York: Guilford Press.

Cablk, M. E., B. Kjerfve, W. K. Michener, J. R. Jensen. 1994. Impacts of Hurricane Hugo on a coastal forest: Assessment using LANDSAT TM data. *Geocarto International* 2:15-24.

Cardona, C., R. Davidson, and C. Villacis. 1999. Understanding urban seismic risk around the world. Pp. 262-263 in *Natural Disaster Management*, J. Ingleton, ed. Leicester, England: Tudor Rose.

Carrara, A., and F. Guzzetti, eds. 1996. *Geographical Information Systems in Assessing Natural Hazards*. Dordrecht, The Netherlands: Kluwer Academic.

Carrara, A., M. Cardinali, R. Detti, F. Guzzetti, V. Pasquia, and P. Reichenbach. 1991. GIS techniques and statistical models in evaluating landslide hazard. *Earth Surface Processes and Landforms* 16:427-445.

CDC (Centers for Disease Control and Prevention). 1995. Heat-related illnesses and deaths—United States 1994-1995. *Mortality and Morbidity Weekly Reports* 44:465-468.

Chakraborty, J., and M. P. Armstrong. 1997. Exploring the use of buffer analysis for the identification of impacted areas in environmental equity assessment. *Cartography and Geographic Information Systems* 24:145-157.

Changnon, S. A., ed. 1996. *The Great Flood of 1993: Causes, Impacts, and Responses*. Boulder, CO: Westview Press.

Changnon, S. A. 1998. The historical struggle with floods in the Mississippi River Basin. *Water International* 23:263-271.

Changnon, S. A., and D. R. Easterling. 2000. U.S. policies pertaining to weather and climate extremes. *Science* 289:2053-2055.

Chien, P. 2000. *Endeavor* maps the world in three dimensions. *GeoWorld* 13(4):32-38.

Chou, Y. H. 1992. Management of wildfires with a geographical information system. *International Journal of Geographical Information Systems* 6:123-140.

Cinti, F. A. 1994. Southern California shakes again: the earthquake of January 17, 1994. *Systema Terra—Remote Sensing and the Earth* 3(2):27-30.

Clark, W. A. V., and K. L. Avery. 1976. The effects of data aggregation in statistical analysis. *Geographical Analysis* 8:428-438.

Coch, N. K. 1995. *Geohazards: Natural and Human.* Englewood Cliffs, NJ: Prentice-Hall.

COES (California Office of Emergency Services). 1994. East Bay Hills Fire Operations Review Group: A Multi-Agency Review of the October 1991 Fire in Oakland/Berkeley Hills. Washington, DC: National Commission on Wildfire Disasters.

Collins, R.F. 1998. Risk Visualization as a Means for Altering Hazard Cognition. Ph.D. dissertation. University of South Carolina.

Colten, C. E. 1986. Industrial wastes in southeast Chicago: Production and disposal 1870-1970. *Environmental Review* 10:92-105.

Colten, C. E. 1991. A historical perspective on industrial wastes and groundwater contamination. *Geographical Review* 81:215-228.

Comfort, L., B. Wisner, S. Cutter, R. Pulwarty, K. Hewitt, A. Oliver-Smith, J. Wiener, M. Fordham, W. Peacock, and F. Krimgold. 1999. Reframing disaster policy: The global evolution of vulnerable communities. *Environmental Hazards* 1:39-44.

Committee on Risk Assessment of Hazardous Air Pollutants, National Research Council. 1994. *Science and Judgment in Risk Assessment.* Washington, DC: National Academy Press.

Cova, T. J., and R. L. Church. 1997. Modeling community evacuation vulnerability using GIS. *International Journal of Geographical Information Science* 11:763-784.

Covello, V. T., and J. Mumpower. 1985. Risk analysis and risk management: An historical perspective. *Risk Analysis* 5:103-120.

Cowen, D. J., and J. R. Jensen. 1998. Extraction and modeling of urban attributes using remote sensing technology. Pp. 164-188 in *People and Pixels: Linking Remote Sensing and Social Science.* D. Liverman, E. F. Moran, R. R. Rindfuss, and P. C. Stern, eds. Washington DC: National Academy Press.

CRED (Center for Research on the Epidemiology of Disasters). 2000. The OFDA/CRED international disaster database. *http://www.cred.be/emdat.*

Curran, E.B., R. L. Holle, and R. E. Lopez. 1997. Lightning Fatalities, Injuries and Damage Reports in the United States, 1959-1994. NOAA Technical Memo No. NWS SR-193. Silver Spring, MD: NOAA.

Cutter, S. L. 1993. *Living with Risk: The Geography of Technological Hazards.* London: Edward Arnold.

Cutter, S. L., ed. 1994. *Environmental Risks and Hazards.* Englewood Cliffs, NJ: Prentice-Hall.

Cutter, S. L. 1995. The forgotten casualties: Women, children, and environmental change. *Global Environmental Change* 5:181-194.

Cutter, S. L. 1996a. Vulnerability to environmental hazards. *Progress in Human Geography* 20:529-539.

Cutter, S. L. 1996b. Societal responses to environmental hazards. *International Social Science Journal* 150:525-536.

Cutter, S., and M. Ji. 1997. Trends in U.S. hazardous materials transportation spills. *Professional Geographer* 49:318-331.

Cutter, S. L., and W. H. Renwick. 1999. *Exploitation, Conservation, Preservation: A Geographic Perspective on Natural Resource Use.* New York: John Wiley & Sons.

Cutter, S. L., D. Holm, and L. Clark. 1996. The role of geographic scale in monitoring environmental justice. *Risk Analysis* 16:517-526.

Cutter, S. L., D. S. K. Thomas, M. E. Cutler, J. T. Mitchell, and M. S. Scott. 1999. *The South Carolina Atlas of Environmental Risks and Hazards.* Columbia, SC: University of South Carolina Press.

Cutter, S. L., J. T. Mitchell, M. S. Scott. 2000. Revealing the vulnerability of people and places: A case study of Georgetown County, South Carolina. *Annals of the Association of American Geographers* 90:713-737.

Cutter, S. L., M. E. Hodgson, and K. Dow. 2001. Subsidized inequities: The spatial patterning of environmental risks and federally assisted housing. *Urban Geography* 22(1):29-53.

Dangermond, J. 1991. Applications of GIS to the international decade for natural hazards reduction. Pp. 445-468 in *Proceedings, Fourth International Conference on Seismic Zonation*. Earthquake Engineering Research Institute, Stanford University, Palo Alto, CA.

Davidson, R. 2000. An Urban Earthquake Disaster Risk Index. *http://pangea.stanford.edu/~tucker/eri/edri.html.*

Davies, J. C., ed. 1996. *Comparing Environmental Risks: Tools for Setting Government Priorities*. Washington, DC: Resources for the Future.

Davis, B. A. 1993. Mission accomplished. *NASA Tech Briefs* 17:14-16.

Davis, M. 1998. *Ecology of Fear*. New York: Metropolitan Books.

Dietz, T., and T. W. Rycroft. 1987. *The Risk Professionals*. New York: Russell Sage Foundation.

DITF (Disaster Information Task Force). 1997. *Harnessing Information and Technology for Disaster Management—The Global Disaster Information Network*. Washington, DC: U.S. Government Printing Office.

Downing, T. E., 1991. Vulnerability to hunger and coping with climate change in Africa. *Global Environmental Change* 1:365-380.

Easterling, D. R., G. A. Meehl, C. Parmesan, S. A. Changnon, T. R. Karl, and L. O. Mearns. 2000. Climate extremes: Observations, modeling, and impacts. *Science* 289: 2068-2074.

EIA (Energy Information Administration). 1998. *Inventory of Power Plants in the United States*. Washington, DC: U.S. Department of Energy.

Enarson, E. P., and B. H. Morrow, eds. 1998. *The Gendered Terrain of Disaster*. New York: Praeger.

Estes, J. E., K. C. McGwire, G. A. Fletcher, and T. W. Foresman. 1987. Coordinating hazardous waste management activities using geographical information systems. *International Journal of Geographical Information Systems* 1:359-377.

Etkin, D. 1999. Risk transference and related trends: Driving forces toward more mega-disasters. *Environmental Hazards* 1:69-75.

FEMA (Federal Emergency Management Agency). 1995. *National Mitigation Strategy*. Washington, DC: FEMA Mitigation Directorate.

FEMA (Federal Emergency Management Agency). 1997a. *Multi Hazard Identification and Risk Assessment*. Washington, DC: U.S. Government Printing Office.

FEMA (Federal Emergency Management Agency). 1997b. *Recovery Times*. Bismark, and North Dakota Department of Emergency Management North Dakota, Issue 1, April 1997.

FEMA (Federal Emergency Management Agency). 1997c. *HAZUS User's Manual*.

FEMA (Federal Emergency Management Agency). 2000a. HAZUS 99 Estimated Annual Earthquake Losses for the United States. FEMA 366. Available at *www.fema.gov/hazus*.

FEMA (Federal Emergency Management Agency). 2000b. Flood Hazard Mapping: FEMA. publishes appendix for Airborne Light Detection and Ranging Systems. *http://www.fema.gov/mit/tsd/mm_lidar.htm*.

FEMA (Federal Emergency Management Agency). 2000c. *Mitigation—National Dam Safety Programs—About Dams. http://www.fema.gov/MIT/damsafe/about.htm.*

FEMA/NIST/NSF/USGS (Federal Emergency Management Agency/National Institute of Standards and Technology/National Science Foundation/U.S. Geological Survey). 1999. NEHRP: Partners in Earthquake Mitigation, Report to Congress Fiscal Years 1997 & 1998. Washington, DC: U.S. Government Printing Office.

FGDC (Federal Geographic Data Committee). 1997. National Spatial Data Infrastructure. *http://www.fgdc.gov/publications/documents/geninfo/fgdc.pdf.*

Fischhoff, B., P. Slovic, S. Lichtenstein, S. Read, and B. Combs. 1978. How safe is safe enough? A psychometric study of attitudes towards technological risks and benefits. *Policy Sciences* 9:127-152.

Fischhoff, B., P. Slovic, and S. Lichtenstein. 1979. Weighing the risks. *Environment* 21(4):17-20, 32-38.

Foote, K. 1997. *Shadowed Ground: America's Landscape of Violence and Tragedy*. Austin: University of Texas Press.

Frankel, A., C. Mueller, T. Barnhard, D. Perkins, E.V. Leyendecker, N. Dickman, S. Hanson, and M. Hopper. 1996. National Seismic Hazard Maps: Documentation. USGS Open-File Report 96-532. Reston, VA: USGS.

Frankel, A., C. Mueller, T. Barnhard, D. Perkins, E.V. Leyendecker, N. Dickman, S. Hanson, and M. Hopper. 1997. National 1996 Seismic Hazard Maps. Open-File Report 97-131. Reston, VA: USGS.

Frerich, R.R. 2000. John Snow Site. *http://www.ph.ucla.edu/epi/snow.html.*

Fujita, T. T. 1987. *U.S. Tornadoes, Part 1: 70-Year Statistics*. Chicago: The University of Chicago.

Garcia, A. W., Jarvinen, B. R., and R. E. Schuck-Kolben. 1990. Storm surge observations and model hindcast comparison for Hurricane Hugo. *Shore and Beach* 58:15-21.

Geohazards International. 2000. RADIUS (Risk Assessment Tools for Diagnosis of Urban Areas against Seismic Disasters). *http://www.geohaz.org/radius.*

Godschalk, D., T. Beatley, P. Berke, D. Brower, and E. Kaiser. 1999. *Natural Hazard Mitigation: Recasting Disaster Policy and Planning*. Washington DC: Island Press.

Golden, J., and J. Snow. 1991. Mitigation against extreme windstorms. *Review of Geophysics* 29:477-504.

Golding, D. 1992. A social and programmatic history of risk research. Pp. 23-52 in *Social Theories of Risk*. S. Krimsky and D. Golding, eds. Westport, CT: Praeger.

Goldman, B. A. 1991. *The Truth About Where You Live: An Atlas for Action on Toxins and Mortality*. New York: Random House.

Goldsteen, R. L., and J. K. Schorr. 1991. *Demanding Democracy After Three Mile Island*. Gainesville, FL: University of Florida Press.

Goodchild, M., and S. Gopal. 1992. *Accuracy of Spatial Databases*. New York: Taylor and Francis.

Gornitz, V. N., R. C. Daniels, T. W. White, and K. R. Birdwell. 1994. The development of a coastal risk assessment database: Vulnerability to sea-level rise in the U.S. southeast. *Journal of Coastal Research* 12:327-338.

Gray, W. M., C. W. Landsea, P. W. Mielke, Jr., and K. J. Berry. 1999. Extended Range Forecast of Atlantic Seasonal Hurricane Activity and US Landfall Strike Probability for 2000. *http://typhoon/atmos.colostate.edu/forecasts/2000/fcst2000.*

Grazulis, T. P. 1993. *Significant Tornadoes, 1680-1991*. St. Johnsbury, VT: Tornado Project of Environmental Films.

Griffin, L. R., and T. L. Rutherford. 1994. Comparison of air dispersion modeling results with ambient air sampling data—A case study at Tacoma landfill, a NPL site. *Environmental Progress* 13:155-162.

Gruntfest, E. 1996. *Twenty Years Later: What We Have Learned Since the Big Thompson Flood*. Special Publication No. 33. Boulder, CO: Natural Hazards Research and Applications Information Center.

GWRMRC (Governor's Wildfire Response and Mitigation Review Committee). 1998. Through the Flames...An Assessment of Florida's Wildfires of 1998. Final Report. *http://www.floridadisaster.org.*

Hamilton, J. T., and W. K. Viscusi. 1999. *Calculating Risks: The Spatial and Political Dimensions of Hazardous Waste Policy.* Cambridge, MA: MIT Press.

Heinz Center for Science, Economics, and the Environment. 2000a. *The Hidden Costs of Coastal Hazards: Implications for Risk Assessment and Mitigation.* Covello, CA: Island Press.

Heinz Center for Science, Economics, and the Environment. 2000b. Evaluation of Erosion Hazards (Summary). Washington DC: The Heinz Center. Full text of the report is available at *http://www.heinzcenter.org.*

Hebert, P., J. Jerrel, and M. Mayfield. 1996. The deadliest, costliest, and most intense United States hurricanes of this century (and other frequently requested hurricane facts). NOAA Technical Memorandum NWS NHC-31 (Feb). Coral Gables, FL: NHC.

Hewitt, K., ed. 1983. *Interpretations of Calamity.* Winchester, MA: Allen and Unwin.

Hewitt, K. 1997. *Regions of Risk: A Geographical Introduction to Disasters.* Essex, UK: Longman.

Hird, J. A. 1994. *Superfund: The Political Economy of Environmental Risk.* Baltimore, MD: Johns Hopkins University Press.

Holle, R. L., R. E. Lopez, L. J. Arnold, and J. Endres. 1996. Insured lightning caused property damage in three western states. *Journal of Applied Meteorology* 35:1344-1351.

Houston, S. H., W. A. Shaffer, M. D. Powell, and J. Chen. 1999. Comparisons of HRD and SLOSH surface wind field in hurricanes: Implications for storm surge modeling. *Weather and Forecasting* 14:671-686.

HPC (Hydrometeorologic Prediction Center). 2000. *http://www.hpc.ncep.noaa.gov/.*

Hutchinson, C.F. 1998. Social science and remote sensing in famine early warning. Pp.189-196 in *People and Pixels: Linking Remote Sensing and Social Science.* D. Liverman, E. F. Moran, R. R. Rindfuss, and P. C. Stern, eds. Washington DC: National Research Council.

IBHS (Institute for Business Home Safety). 1998. *The Insured Cost of Natural Disasters: A Report on the IBHS Paid Loss Database.* Boston, MA: IBHS.

IFMRC (Interagency Floodplain Management Review Committee). 1994. *Sharing the Challenge: Floodplain Management into the 21st Century.* Washington, DC: Administration Floodplain Management Task Force.

IFRCRCS (International Federation of Red Cross and Red Crescent Societies). 1998. *World Disasters Report.* Oxford, UK: Oxford University Press.

IIPLR (Insurance Institute for Property Loss Reduction and Insurance Research Council). 1995. *Coastal Exposure and Community Protection: Hurricane Andrew's Legacy.* Boston, MA and Wheaton, IL: IIPLR and Insurance Research Council.

IJC (International Joint Commission). 2000. *Living with the Red.* Washington, DC: IJC.

Ingleton, J., ed. 1999. *Natural Disaster Management.* Leicester, England: Tudor Rose.

ISO (Insurance Services Office, Inc.). 2000. A Half Century of Hurricane Experience. *http://www.iso.com/studies_analyses/hurricane_experience/index.html.*

Jelesnianski, C. P., J. Chen, and W. A. Shaffer. 1992. SLOSH: Sea, Lake, and Overland Surges from Hurricanes. NOAA Technical Report NWS 48. Washington, DC: U.S. Department of Commerce.

Jensen, J. R. 1996. *Introductory Digital Image Processing: A Remote Sensing Perspective 2nd ed.* Saddle River, NJ: Prentice-Hall.

Jensen, J. R. 2000. *Remote Sensing of the Environment: An Earth Resource Perspective.* Upper Saddle River, NJ: Prentice-Hall.

Jensen, J. R., and D. J. Cowen. 1999. Remote sensing of urban/suburban infrastructure and socio-economic attributes. *Photogrammetric Engineering and Remote Sensing* 65:611-622.

Jensen, J. R., S. Narumalani, O. Weatherbee, M. Murday, W. Sexton, and C. Green. 1993. Coastal environmental sensitivity mapping for oil spills in the United Arab Emirates using remote sensing and GIS technology. *Geocarto International* 2:5-13.

Jensen, J. R., J. N. Halls, and J. Michel. 1998. A systems approach to environmental Sensitivity Index (ESI) mapping for oil spill contingency planning and response. *Photogrammetric Engineering and Remote Sensing* 64:1003-1014.

Johnson, G. O. 1992. GIS applications in emergency management. *URISA Journal* 4(1):66-72.

Kasperson, J. X., R. E. Kasperson, and B. L. Turner III, eds. 1995. *Regions at Risk: Comparisons of Threatened Environments*. Tokyo: United Nations University Press.

Kates, R.W. 1971. Natural hazard in human ecological perspective: Hypotheses and models. *Economic Geography* 47:438-451.

Kates, R.W. 1978. *Risk Assessment of Environmental Hazard*. SCOPE 8. New York: John Wiley & Sons.

Kerr, R. A. 1990. Hurricane forecasting shows promise. *Science* 247:917.

Kilbourne, E. 1989. Heat waves. *The Public Health Consequences of Disasters 1989*. CDC Monograph. Atlanta, GA: U.S. Department of Health and Human Services, Public Health Service, Centers for Disease Control and Prevention.

King, M. D., and D. D. Herring. 2000. Monitoring Earth's vital signs. *Scientific American* 282(4) (April):92-97.

Kirn, W. 2000. Backyard infernos. *Time* (August 21):70-71.

Knutson, C. 1997. A Comparison of Droughts, Floods, and Hurricanes in the U.S. National Drought Mitigation Center. Available at *http://enso.unl.edu/ndmc/impacts/compare.htm*.

Krimsky, S., and D. Golding, eds. 1992. *Social Theories of Risk*. Westport, CT: Praeger.

Kunreuther, H., and P. Slovic, eds. 1996. Challenges in risk assessment and risk management. Special Issue, *Annals of the American Academy of Political and Social Science* 545:8-183.

Kunreuther, H., and R. J. Roth, Sr., eds. 1998. *Paying the Price: The Status and Role of Insurance Against Natural Disasters in the United States*. Washington DC: Joseph Henry Press.

LaRoe, E. T., G. S. Farris, C. E. Puckett, P. D. Doran, and M. J. Mac, eds. 1995. *Our Living Resources: A Report to the Nation on the Distribution, Abundance, and Health of U.S. Plants, Animals, and Ecosystems*. Washington, DC: U.S. Department of the Interior, National Biological Survey.

Larson, E. 1999. *Isaac's Storm: A Man, A Time, and the Deadliest Hurricane in History*. New York: Crown.

Lillibridge, S. R. 1997. Tornadoes. Pp. 220-244 in *The Public Health Consequences of Disasters*. E. K. Noji, ed. New York: Oxford University Press.

Lillisand, T. M., and R. W. Kiefer. 1994. *Remote Sensing and Image Interpretation*. New York: John Wiley & Sons.

Liverman, D. 1990. Drought in Mexico: Climate, agriculture, technology and land tenure in Sonora and Puebla. *Annals of the Association of American Geographers* 80:49-72.

Lott, N. 1993. The Big One! A Review of the March 12-14, 1993 "Storm of the Century." Technical Report 93-01. Asheville, NC: NCDC (National Climatic Data Center) Research Customer Service Group.

Lott, N., and M. Sittel. ND. *The February 1994 Ice Storm in the Southeastern U.S.* Asheville, NC: NCDC (National Climatic Data Center).

Lowry, J. H., H. J. Miller, and G. F. Hepner. 1995. A GIS-based sensitivity analysis of community vulnerability to hazardous contaminants on the Mexico/US border. *Photogrammetric Engineering and Remote Sensing* 61:1347-1359.

L. R. Johnson Associates. 1992. Floodplain Management in the United States: An Assessment Report. FIA-18. Washington, DC: Federal Interagency Floodplain Management Task Force.

Lynn, F., and J. Kartez. 1994. Environmental democracy in action: The Toxic Release Inventory. *Environmental Management* 18:511-521.

Malmquist, D., and R. Murnane. 1999. A working example of a public-private partnership. Pp. 271-274 in J. Ingleton, ed. *Natural Disaster Management.* Leicester, England: Tudor Rose.

Mangan, D. 1999. Wildland Fire Fatalities in the United States, 1990-1998. Technical Report 9951-2802-MTDC, March. Missoula, MT: Missoula Technology and Development Center.

Marcello, B. 1995. FEMA's new GIS reforms emergency response efforts. *GIS World* 8(12):70-73.

Mason, R. R., Jr., and B. A. Weiger. 1995. Stream Gauging and Flood Forecasting: A Partnership of the U.S. Geological Survey and the National Weather Service. USGS Factsheet 209-95. *http://water.usgs.gov/wid/FX-209-95/mason-weiger.html.*

McCullough, D. 1968. *The Johnstown Flood.* New York: Simon & Schuster.

Meija-Navarro, M,. and E. E. Wohl. 1994. Geologic hazard and risk evaluation using GIS: Methodology and model applied to Medellin, Colombia. *Bulletin of the Association of Engineering Geologists* 31:459-481.

Mileti, D. 1980. Human adjustment to the risk of environmental extremes. *Sociology and Social Research* 64:327-347.

Mileti, D. 1999. *Disasters by Design: A Reassessment of Natural Hazards in the United States.* Washington DC: Joseph Henry Press.

Mitchell, J. K., N. Devine, and K. Jagger. 1989. A contextual model of natural hazard. *Geographical Review* 79:391-409.

Mitchell, J. T., M. S. Scott, D. S. K. Thomas, M. Cutler, P. D. Putnam, R. F. Collins, and S. L. Cutter, 1997. Mitigating against disaster: Assessing hazard vulnerability at the local Level, Pp. 563-571 in *GIS/LIS'97 Proceedings.* Bethesda, MD: American Congress on Surveying and Mapping, American Society of Photogrammetry and Remote Sensing, Automated Mapping/Facilities Management, Association of American Geographers, Urban and Regional Information Systems Association.

Monmonier, M. 1991. *How to Lie with Maps.* Chicago, IL: University of Chicago Press.

Monmonier, M. 1997. *Cartographies of Danger: Mapping Hazards in America.* Chicago, IL: University of Chicago Press

Monmonier, M. 1999. *Air Apparent: How Meteorologists Learned to Map, Predict, and Dramatize Weather.* Chicago, IL: University of Chicago Press.

Morgan, M. G., H. K. Florig, M. L. DeKay, and P. Fishbeck. 2000. Categorizing risks for risk ranking. *Risk Analysis* 20:49-58.

Munich Insurance Group. 1998. World Map of Natural Disasters. Available at *http://www.munichre.com/themes/1998/themes6_eng.html.*

Munich Re. 2000. *Topics 2000. Naturkastrophen-Stand der Dinge.* Munich: Munich Re Group.

NAPA (National Academy of Public Administration). 1998. *Geographic Information for the 21st Century—Building a Strategy for the Nation.* Washington, DC: NAPA.

NAPA (National Academy of Public Administration). 1999. Legal Limits on Access to and Disclosure of Disaster Information. Summary Report. Washington, DC: NAPA.

Nash, J. R. 1976. *Darkest Hours.* Chicago, IL: Nelson-Hall.

NCDC (National Climatic Data Center). 1979. Description of the Red River tornado outbreak. *Storm Data and Unusual Weather Phenomena*. April.

NCDC (National Climatic Data Center). 1991. Description of the severe Michigan thunderstorms and damage. *Storm Data and Unusual Weather Phenomena*. March 27, 1991.

NCDC (National Climatic Data Center). 1995. Description of the severe New York thunderstorms and damage. *Storm Data and Unusual Weather Phenomena*. May.

NCDC (National Climatic Data Center). 2000. *Billion Dollar U.S. Weather Disasters*. *http://www.ncdc.noaa.gov/ol/reports/billionz.html*.

NEIC (National Earthquake Information Center). 1999. Earthquake Hazard Program. *http://neic.usgs.gov*.

Newsome, D. E., and J. E. Mitrani. 1993. Geographic information system applications in emergency management. *Journal of Contingencies and Crisis Management* 1:199-202.

NIFC (National Interagency Fire Center). 2000. Fire Statistics. *http://www.nifc.gov/stats/*.

Nishenko, S. 1999. Natural Hazards Exposure Mapping in the United States. United States - Japan Cooperative Program on Natural Resources. Workshop on Seismic Information Systems, Tsukuba Japan. Mimeo.

NLSI (National Lightning Safety Institute). 2000. National Lightning Safety Institute. *http://www.lightningsafety.com/*.

NOAA (National Oceanic and Atmospheric Administration). 1994. The Great Flood of 1993. National Disaster Survey Report. Washington DC: U.S. Department of Commerce.

NOAA (National Oceanic and Atmospheric Administration). 1995. The July 1995 Heat Wave Natural Disaster Survey Report. Silver Spring, MD: U.S. Department of Commerce.

NOAA (National Oceanic and Atmospheric Administration), 2000a. Hurricane Storm Surge Forecasting. *http://tgsv5.nws.noaa.gov/tdl/marine/hursurge*.

NOAA (National Oceanic and Atmospheric Administration). 2000b. Questions and Answers about Lightning. *http://www.nssl.noaa.gov/edu/ltg/*.

NOAA (National Oceanic and Atmospheric Administration) Coastal Services Center. Nda *Alabama Coastal Hazards Assessment*. Charleston, SC: NOAA Coastal Services Center, CD-ROM.

NOAA (National Oceanic and Atmospheric Administration) Coastal Services Center. Ndb. *Community Vulnerability Assessment Tool*. Charleston, SC: NOAA Coastal Services Center, CD-ROM.

NODAK. 2000. Volcano World. *http://volcano.und.nodak.edu*.

NRC (National Research Council). 1983. *Risk Assessment in the Federal Government: Managing the Process*. Washington, DC: National Academy Press.

NRC (National Research Council). 1994. *Science and Judgment in Risk Assessment*. Washington, DC: National Academy Press.

NRC (National Research Council). 1996. *Understanding Risk: Informing Decisions in a Democratic Society*. Washington, DC: National Academy Press.

NRC (National Research Council). 1999a. *Reducing Disaster Losses Through Better Information*. Washington, DC: National Academy Press.

NRC (National Research Council). 1999b. *The Impacts of Natural Disasters: Framework for Loss Estimation*. Washington, DC: National Academy Press.

NWS (National Weather Service). 1994. *Thunderstorms and Lightning: The Underrated Killers*. Washington, DC: U.S. Department of Commerce.

NWS (National Weather Service). 2000. National Weather Service Southern Region Headquarters, Norman, Oklahoma Forecast Office. *http://www.srh.noaa.gov/oun/storms*.

OAS (Organization of American States). 1991. *Primer on Natural Hazard Management in Integrated Regional Development Planning.* Washington DC: Department of Regional Development and Environment, Executive Secretariat for Economic and Social Affairs, OAS.

OAS (Organization of American States). 2000. Caribbean Disaster Mitigation Project. *www.oas.org/cdmp.*

Okazaki, K. 1999. The RADIUS initiative. Pp. 298-301 in *Natural Disaster Management.* J. Ingleton, ed. Leicester, England: Tudor Rose.

Palm, R. I. 1990. *Natural Hazards: An Integrative Framework for Research and Planning.* Baltimore, MD: Johns Hopkins University Press.

Palm, R. 1995. *Earthquake Insurance: A Longitudinal Study of California Homeowners.* Boulder, CO: Westview Press.

Palm, R. I., and M. E. Hodgson. 1992a. *After a California Earthquake: Attitude and Behavior Change.* Chicago, IL: University of Chicago Press.

Palm, R. I., and M. E. Hodgson. 1992b. Earthquake insurance: Mandated disclosure and homeowner response in California. *Annals of the Association of American Geographers* 82:207-222.

Parfit, M. 1998. Living with natural hazards. *National Geographic* Vol. 194 (July):2-39.

Perrow, C. 1999. *Normal Accidents: Living with High-Risk Technologies.* Princeton, NJ: Princeton University Press.

Petak, W. J., and A. A. Atkisson. 1982. *Natural Hazard Risk Assessment and Public Policy.* New York: Springer-Verlag.

Pielke, R.A., Jr. 1997. Reframing the U.S. hurricane problem. *Society and Natural Resources* 10:485-499.

Pielke, R. A. 1999. Hurricane forecasting. *Science* 284:1123.

Pielke, R. A., Jr., and C. W. Landsea. 1998. Normalized hurricane damages in the United States: 1925-1995. *Weather and Forecasting* 13:621-631.

Pielke, R. A., Jr., and C. W. Landsea. 1999. La Niña, El Niño, and Atlantic hurricane damages in the United States. *Bulletin of the American Meteorological Society* 80: 2027-2033.

Pielke, R. A., Jr., and R. A. Pielke, Sr. 1997. *Hurricanes: Their Nature and Impacts on Society.* New York: John Wiley & Sons.

Platt, R. 1995. Lifelines: An emergency management priority for the United States in the 1990s. *Disasters* 15:172-176.

Platt, R. H. 1999. *Disasters and Democracy: The Politics of Extreme Natural Events.* Washington, DC: Island Press.

Presidential/Congressional Commission on Risk Assessment and Risk Management. 1997. *Risk Assessment and Risk Management in Regulatory Decision Making. Vols. 1 and 2.* Washington, DC: U.S. Government Printing Office.

Press, F., and R. M. Hamilton. 1999. Mitigating natural disasters. *Science* 284:1927.

Preuss, J., and G. T. Hebenstreit. 1991. Integrated hazard assessment for a coastal community: Grays Harbor. USGS Open-File Report 91-441-M. Washington, DC: USGS.

Pyne, S. J. 1997. *Fire in America: A Cultural History of Wildland and Rural Fire.* Seattle, WA: University of Washington Press.

Quarantelli, E. L. 1988. Disaster studies: An analysis of the social historical factors affecting the development of research in the area. *International Journal of Mass Emergencies and Disasters* 5:285-310.

Quarantelli, E. L., ed. 1998. *What Is a Disaster? Perspectives on the Question.* London: Routledge.

Radke, J., T. Cova, M. F. Sheridan, A. Troy, L. Mu, and R. Johnson. 2000. Application challenges for geographic information science: Implications for research, education, and policy for emergency preparedness and response. *URISA Journal* 12(2):15-30.

Rappaport, E. N. 2000. Loss of life in the United States associated with recent Atlantic tropical cyclones. *Bulletin of the American Meteorological Society* 81:2065-2073.

Riebsame, W., S. Changnon, and T. Karl. 1991. *Drought and Natural Resources Management in the United States: Impacts and Implications of the 1987-1989 Drought.* Boulder, CO: Westview Press.

Robinson, A., R. Sale, and J. Morrison. 1978. *Elements of Cartography, 4th ed.* New York: John Wiley & Sons.

Saarinen, T. F. 1966. *Perception of the Drought Hazard on the Great Plains.* Chicago, IL: Department of Geography Research Paper No. 106, University of Chicago.

Schmidlin, T. W., and P. S. King. 1995. Risk factors for death in the 27 March 1994 Georgia and Alabama tornadoes. *Disasters* 19:170-177.

Schmidlin, T. W., and P. S. King. 1997. Risk Factors for death in the 1 March 1997 Arkansas Tornadoes. Final Report No. 98 of a Quick Response Grant. Boulder, CO: Natural Hazards Research Applications and Information Center.

Schneider, J., G. Rao, S. Daneshvaran, and J. Perez. 1999. Mitigating property and business losses. Pp. 254-256 in *Natural Disaster Management.* J. Ingleton, ed. Leicester, England: Tudor Rose.

Shinozuka, M., A. Rose, and R. T. Eguchi, eds. 1998. *Engineering and Socioeconomic Impacts of Earthquakes: An Analysis of Electricity Lifeline Disruptions in the New Madrid Area.* Monograph No. 2. Buffalo, NY: Multidisciplinary Center for Earthquake Engineering Research.

Sills, D. L., C. P. Wolf, and V. B. Shelanski, eds. 1982. *Accident at Three Mile Island: The Human Dimension.* Boulder, CO: Westview Press.

Simpson, S. 2000. Raging rivers of rock. *Scientific American* Vol. 283 (July): 24-25.

Sims, J. H., and D. D. Baumann. 1972. The tornado threat: Coping styles of the North and South. *Science* 176:1386-1392.

Slovic, P. 1987. Perception of risk. *Science* 236:280-285.

Slovic, P. 2001. *The Perception of Risk.* London: Earthscan.

Smithsonian Institution, Global Volcanism Program. 1999. Volcanoes of the World, Volcano Basic Data. *http://nmnhgoph.si.edu/gvp/volcano/vbd_alph.htm.*

Starr, C. 1969. Social benefit versus technological risk. *Science* 165:1232-1238.

State of Florida. 2000. Hazard Idenfication and Vulnerability Assessment. *www.dca.state.fl.us/fhcd/programs/ltr/lms/lms_hiva.htm.*

Stewart, J. C., K. L. Martin, and A. R. Jennetta. 1993. U.S. Department of Energy uses GIS to evaluate waste management alternatives. *Geo Info Systems* 3:60-63.

Stewart, S. W. 1977. Real-time detection and location of local seismic events in central California. *Bulletin of the Seismological Society of America* 62:433-452.

Summerhays, J. 1991. Evaluation of risks from urban air pollution in the southeast Chicago area. *Journal of the Air Waste Management Association* 41:884-850.

Suter, G. W., II. 2000. Generic assessment endpoints are needed for ecological risk assessment. *Risk Analysis* 20:173-178.

Timmreck, T. C. 1998. *An Introduction to Epidemiology, 2nd Ed.* Sudbury, MA: Jones and Bartlett.

Toppozada, T., G. Borchardt, W. Haydon, M. Peterson, R. Olson, H. Lagorio, and T. Anvik. 1995. Planning Scenario in Humboldt and Del Norte Counties, California, for a Great Earthquake on the Cascadia Subduction Zone. Sacramento, CA: California Department of Conservation, Division of Mines and Geology. Special Publication 115.

Tucker, C. 1998. Natural Hazards of North America. *National Geographic* 194, (July) Map Supplement.

Ullmann, O. 2000. Facing Mother Nature's fury. *USA Today.* (July 24). 6A.

UNEP (United Nations Environment Program). 1993. *Environmental Data Report 1993-94*. Cambridge, MA: Blackwell.

U.S. Country Studies Program. 1999. *Climate Change: Mitigation, Vulnerability, and Adaptation in Developing and Transition Countries*. Washington, DC: U.S. Department of Energy.

USDOT (U.S. Department of Transportation). 2000. About the Hazardous Material Information System. *http://hazmat.dot.gov/abhmis.htm*.

USEPA (U.S. Environmental Protection Agency). 1987. *Unfinished Business: A Comparative Assessment of Environmental Problems*. Washington, DC: USEPA.

USEPA (U.S. Environmental Protection Agency). 1995. *User's Guide for the Industrial Source Complex (ISC3) Dispersion Models*. IPA-454/B-95-003a. Washington, DC: USEPA.

USEPA (U.S. Environmental Protection Agency). 2000a. About Aloha. *http://response.restoration.noaa.gov/cameo/aloha/html*.

USEPA (U.S. Environmental Protection Agency). 2000b. Cameo. *http://response.restoration.noaa.gov/cameo/intro.html*.

USEPA (U.S. Environmental Protection Agency). 2000c. Guidelines for Ecological Risk Assessment. *http://www.epa.gov/ncea/ecorsk.htm*.

USEPA (U.S. Environmental Protection Agency). 2000d. Comparative Risk Assessment. *http://www.epa.gov/grtlakes/seahome/comrisk.html*.

USEPA (U.S. Environmental Protection Agency). 2000e. Landview III on CD-ROM. *http://www.census.gov/geo/www/tiger/lv3desc.html*.

USEPA (U.S. Environmental Protection Agency). 2000f. Superfund. *http://www.epa.gov/superfund/index.htm*.

USGS (U.S. Geological Survey). 1999. *Digital Atlas of Central America*. Washington, DC: Center for Integration of Natural Disaster Information.

USGS (U.S. Geological Survey). 2000a. Real-Time Water Data. *http://water.usgs.gov/realtime.html*.

USGS (U.S. Geological Survey). 2000b. Earthquake Hazards Program. *http://quake.wr.usgs.gov*.

USGS (U.S. Geological Survey). 2000c. Geologic Hazards. *http://geohazards.cr.usgs.gov*.

USNRC (U.S. Nuclear Regulatory Commission). 1996. Historical Data Summary of the Systematic Assessment of Licensee Performance. NUREG-1214, Rev. 14. Washington, DC: USNRC.

USNRC (U.S. Nuclear Regulatory Commission). 1999. NRC SALP (Systematic Assessment of Licensee Performance) Reports. *http://www.nrc.gov/RIII/rjs2/salpndex.htm*.

USNRC (U.S. Nuclear Regulatory Commission). 2000. Nuclear Reactors. *http://www.nrc.gov/NRC/reactors.html*.

van der Wink, G., R. M. Allen, J. Chapin, M. Crooks, W. Fraley, J. Krantz, A. M. Lavigne A., LeCuyer, E. K. MacColl, W. J. Morgan, B. Ries, E. Robinson, K. Rodriquez, M. Smith, and K. Sponberg. 1998. Why the United States is becoming more vulnerable to natural disasters. *EOS*, Transactions, American Geophysical Union 79(44): 533, 537.

Wagman, D. 1997. Fires, hurricanes prove no match for GIS. *Earth Observation Magazine* 6(2):27-29.

Warrick, R. A. 1975. *Drought Hazard in the United States: A Research Assessment*. Boulder, CO: University of Colorado.

Watson Technical Consulting. 2000. Watson Technical Consulting. *http://www.methaz.com*.

Welch, R. 1982. Spatial resolution requirements for urban studies. *International Journal of Remote Sensing* 3:139-146.

White, G. F. 1964. Choice of Adjustment to Floods. Chicago, IL: University of Chicago, Department of Geography Research Paper No. 93.

White, G. F., ed. 1974. *Natural Hazards: Local, National, Global.* New York: Oxford University Press.

White, G. F. 1988. Paths to risk analysis. *Risk Analysis* 8:171-175.

White, G. F. 1994. A perspective on reducing losses from natural hazards. *Bulletin of the American Meteorological Society* 75:1237-1240.

White, G. F., and J. E. Haas. 1975. *Assessment of Research on Natural Hazards.* Cambridge, MA: MIT Press.

Whyte, A. V. T., and I. Burton. 1980. *Environmental Risk Assessment.* SCOPE 15. New York: John Wiley & Sons.

Wilhite, D. A., ed. 1993. *Drought Assessment, Management, and Planning: Theory and Case Studies.* Dordrecht, The Netherlands: Kluwer Academic.

Wilhite, D. A. 1996. A methodology for drought preparedness. *Natural Hazards* 13:229-252.

Wilhite, D. A. 1997. Responding to drought: Common threads from the past, visions for the future. *Journal of the American Water Resources Association* 33:951-959.

World Economic Forum. 2000. *Pilot Environmental Sustainability Index.* New Haven, CT: Yale Center for Environmental Law and Policy.

Wysession, M. E. 1996. How well do we utilize global seismicity? *Bulletin of the Seismological Society of America* 86(5):1207-1209.

Yeats, R. S., K. Sieh, and C. R. Allen. 1997. *The Geology of Earthquakes.* New York: Oxford University Press.

Zack, J. A., and R. A. Minnich. 1991. Integration of a geographic information system with a diagnostic wind field model for fire management. *Forest Science* 37:560-573.

Appendixes

APPENDIX A

Selected Hazard Assessment Models

A list of several modeling software packages is provided below. User guides, source codes, and examples using the models can be downloaded from the Web sites of the agency that developed the model. Models are updated and revised regularly and Internet addresses are often updated as well. Each agency also has a search engine as part of its home page, which should assist in locating this information as such changes are made.

TABLE A-1 Sources and Targets of Hazard Assessment Models

Model	Targeted Hazard	Reference/Source
SCREEN3D	Airborne toxic plume dispersion	U.S. Environmental Protection Agency (USEPA) *http://www.epa.gov/scram001/t23.htm*
Industrial Source Complex (ISC3)	Airborne toxic plume dispersion	USEPA *http://www.epa.gov/scram001/t23.htm*
Industrial Waste Air Model (IWAIR)	Airborne toxic health risks	USEPA *http://www.epa.gov/epaoswer/nonhw/industd/air/user.pdf*
Areal Locations of Hazardous Atmosphere (ALOHA)	Airborne toxic plume dispersion	USEPA *http://response.restoration.noaa.gov/cameo/aloha.html*
Urban Airshed Model (UAM)	Air quality (industrial emissions)	USEPA *http://www.epa.gov/scram001/t23.htm*
HAPEM	Hazardous air pollutant exposure	USEPA *http://www.epa.gov/AMD/hazair.html*
Mobile6	Vehicle emissions	USEPA *http://www.epa.gov/otaq/m6.htm*
AVN	Hurricane track forecast	National Oceanic and Atmospheric Administration (NOAA) *http://www.nhc.noaa.gov/aboutmodels.html*
NOGAPS	Hurricane track forecast	NOAA *http://www.nhc.noaa.gov/aboutmodels.html*
UKMET	Hurricane track forecast	NOAA *http://www.nhc.noaa.gov/aboutmodels.html*
GFDL	Hurricane track forecast	NOAA *http://www.nhc.noaa.gov/aboutmodels.html*
GFDI	Hurricane track forecast	NOAA *http://www.nhc.noaa.gov/aboutmodels.html*

LBAR	Hurricane track forecast	NOAA *http://www.nhc.noaa.gov/aboutmodels.html*
CLIPER	Hurricane track forecast	NOAA *http://www.nhc.noaa.gov/aboutmodels.html*
BAM	Hurricane track forecast	NOAA *http://www.nhc.noaa.gov/aboutmodels.html*
NHC90/NHC91	Hurricane track forecast	NOAA *http://www.nhc.noaa.gov/aboutmodels.html*
SHIFOR	Hurricane intensity forecast	NOAA *http://www.nhc.noaa.gov/aboutmodels.html*
SHIPS	Hurricane intensity forecast	NOAA *http://www.nhc.noaa.gov/aboutmodels.html*
GFDL	Hurricane intensity forecast	NOAA *http://www.nhc.noaa.gov/aboutmodels.html*
GFDI	Hurricane intensity forecast	NOAA *http://www.nhc.noaa.gov/aboutmodels.html*
TTSURGE and SURGE	Storm surges of landfalling hurricanes	Federal Emergency Management Agency (FEMA) *http://www.fema.gov/mit/tsd/EN_coast.htm*
SLOSH	Storm surge inundation zones based on "typical" hurricanes	Jelesnianski et al. (1992) (with annual updates)
UVA/CERC Storm Risk	Coastal erosion	Anders et al. (1989), Dolan and Davis (1994)
Storm-induced BEAch Change (SBEACH)	Coastal erosion	U.S. Army Corps of Engineers *http://chl.wes.army.mil/software/aces/sbeach*
Coastal Erosion Information System	Coastal erosion	Dolan and Kimball (1988), Gornitz et al. (1994)

continues

TABLE A-1 Continued

Model	Targeted Hazard	Reference/Source
Water Erosion Prediction Project (WEPP)	Soil loss and runoff	U.S. Department of Agriculture *http://topsoil.nserl.purdue.edu/nserlweb/weppmain/weppdocs.html*
Distributed Rainfall-Runoff Model (DR3M)	Rainfall-runoff	U.S. Geological Survey (USGS) *http://water.usgs.gov/software/dr3m.html*
Precipitation-Runoff Modeling System (PRMS)	Rainfall-runoff	USGS *http://water.usgs.gov/software/prms.html*
Watershed Analysis and Detention Design Model (HEC-1)	Rainfall-runoff with application to reservoir and dam design parameters	U.S. Army Corps of Engineers— Hydrologic Engineering Center *http://www.hec.usace.army.mil/software/*
Flood Flow Frequency	Flooding	Interagency Advisory Committee on Water Data (1982)
Flood Forecasting	Flooding	Interagency Advisory Committee on Water Data (1982)
Eta, NGM, AVN, meso-Eta, and RUC	Flooding—quantitative precipitation forecasts	Hydrometeorological Prediction Center and National Weather Service *http://www.hpc.ncep.noaa.gov/html/fcst2.html*
Flood Damage Analysis package (HEC-FDA)	Social cost of flood events—potential damage	U.S. Army Corps of Engineers— Hydrologic Engineering Center *http://www.hec.usace.army.mil/software/*
Interior Flood Hydrology Model (HEC-IFH)	Levee system design	U.S. Army Corps of Engineers— Hydrologic Engineering Center *http://www.hec.usace.army.mil/software/*
DAMBRK	Dam-break flood forecasting	National Weather Service *http://hsp.nws.noaa.gov/oh/hrl/rvrmech/rvrmain.htm*

BREACH	Earthen dam breaches	National Weather Service *http://hsp.nws.noaa.gov/oh/hrl/rvrmech/rvrmain.htm*
Enhanced Stream Water Quality Model (QUAL2E)	Water quality	USEPA *http://www.epa.gov/QUAL2E_WINDOWS/*
HEC5-Q	Water quality	U.S. Army Corps of Engineers— Hydrologic Engineering Center *http://www.hec.usace.army.mil/software/*
Agricultural Non-Point Source Pollution Model (AGNPS)	Water quality	U.S. Department of Agriculture *http://www.wcc.nrcs.usda.gov/water/factsheets/agnps.html*
Better Assessment Science Integrating Point and Non-point Sources (BASINS)	Water quality	USEPA *http://www.epa.gov/OST/BASINS/basinsv2.htm*
HEC-6 and HEC-18	Sediment transport	U.S. Army Corps of Engineers *http://www.hec.usace.army.mil/software/*
Percent of Normal	Drought	Willeke et al. (1994)
Deciles	Drought	Gibbs and Maher (1967)
Palmer Drought Severity Index	Drought	Palmer (1965), Alley (1984)
Standardized Precipitation Index	Drought	McKee et al. (1993)
Crop Moisture Index	Drought	Palmer (1968)

continues

TABLE A-1 Continued

Model	Targeted Hazard	Reference/Source
Surface Water Supply Index	Drought	Shafer and Dezman (1982), Doesken et al. (1991)
National Rainfall Index	Drought	Gommes and Petrassi (1994)
Dependable Rains	Drought	Le Houérou et al. (1993)
Keetch-Byram Drought Index	Wildfire	Keetch and Byram (1968)
Haines Index	Wildfire	Haines (1988)
National Fire Danger Rating System	Wildfire	Burgan (1988), Deeming et al. (1977)
WHIMS	Wildfire hazard information	Boulder County, Colorado *http://www.boco.co.gov/gislu/whims.html*
RISK	Earthquake	Bender and Perkins (1987)
NEHRP Effective Peak Acceleration	Earthquake	Algermissen et al. (1982), Frankel et al. (1997a, b)
NEHRP Effective Peak Velocity	Earthquake	Algermissen et al. (1982)
Heat/Health Index	Heat/health warning—watch system	Kalkstein et al. (1996)
Consequences Assessment Tool Set (CATS)	Disaster analysis	FEMA (1996) *http://www.saic.com/products/software/cats/cats.html*

REFERENCES

Algermissen, S.T., D. M. Perkins, P. C. Thenhaus, and B. L. Bender. 1982. Probabilistic Estimates of Maximum Acceleration and Velocity in Rock in the Contiguous United States. U.S. Geological Survey, Open File Report 82-1033. Reston, VA: USGS.

Alley, W. M. 1984. The Palmer Drought Severity Index: Limitations and assumptions. *Journal of Climate and Applied Meteorology* 23:1100-1109.

Anders, F., S. Kimball, and R. Dolan. 1989. *Coastal Hazards: National Atlas of the United States.* Map. Reston, VA: U.S. Geological Survey.

Bender, B., and D. Perkins. 1987. SEISRISK III: A Computer Program for Seismic Hazard Estimation. U.S. Geological Survey Bulletin 1772. Reston, VA: USGS.

Burgan, R. E. 1988. 1988 Revisions to the 1978 National Fire-Danger Rating System. Research Paper SE-273. Asheville, NC: U.S. Department of Agriculture, Forest Service, Southeastern Forest Experiment Station.

Deeming, J. E., R. E. Burgan, and J. D. Cohen. 1977. The National Fire-Danger Rating System— 1978. Gen. Tech. Rep. INT-39. Ogden, UT: U.S. Department of Agriculture, Forest Service, Intermountain Forest and Range Experiment Station.

Doesken, N. J., T. B. McKee, and J. Kleist. 1991. Development of a Surface Water Supply Index for the Western United States. Climatology Report No. 91-3. Fort Collins, CO: Colorado State University.

Dolan, R., and R. E. Davis. 1994. Coastal storm hazards. *Journal of Coastal Research, Special Issue No. 12, Coastal Hazards: Perception, Susceptibility and Mitigation.* Coastal Education and Research Foundation, Inc. Pps. 103-114. Charlottesville, VA: The Coastal Education and Research Foundation.

Dolan, R., and S. Kimball. 1988. *Coastal Erosion and Accretion: National Atlas of the United States.* Reston, VA: USGS.

FEMA (Federal Emergency Management Agency). 1996. *Consequence Assessment Tool Set and Operations Concept.* Washington, DC: FEMA Protective Services, Information Technology Services.

Frankel, A., C. Mueller, T. Barnhard, D. Perkins, E. V. Leyendecker, N. Dickman, S. Hanson, and M. Hopper. 1997a. California/Nevada 1996 Seismic Hazard Maps. U.S. Geological Survey, Open-File Report 97-130. Reston, VA: USGS.

Frankel, A., C. Mueller, T. Barnhard, D. Perkins, E. V. Leyendecker, N. Dickman, S. Hanson, and M. Hopper. 1997b. National 1996 Seismic Hazard Maps. U.S. Geological Survey, Open-File Report 97-131. Reston, VA: USGS.

Gibbs, W. J., and J. V. Maher. 1967. Rainfall Deciles as Drought Indicators. Bureau of Meteorology Bulletin No. 48. Melbourne, Australia: Bureau of Meteorology.

Gommes, R., and F. Petrassi. 1994. Rainfall Variability and Drought in Sub-Saharan Africa Since 1960. Agrometeorology Series Working Paper No. 9. Rome: Food and Agriculture Organization.

Gornitz, V. N., R. C. Daniels, T. W. White, and K. R. Birdwell. 1994. The development of a coastal risk assessment database: Vulnerability to sea-level rise in the U.S. Southeast. *Journal of Coastal Research* 12:327-338.

Haines, D. A. 1988. A lower atmospheric severity index for wildland fire. *National Weather Digest* 13(2):23-27.

Interagency Advisory Committee on Water Data. 1982. Guidelines for Determining Flood Flow Frequency. Bulletin No. 17B of the Hydrology Subcommittee, U.S. Department of Interior. Reston, VA: USGS, Office of Water Data Coordination.

Jelesnianski, C. P., J. Chen, and W. A. Shaffer. 1992. SLOSH: Sea, Lake, and Overland Surges from Hurricanes. Technical Report NWS 48. Washington, DC: U.S. Department of Commerce, National Oceanic and Atmospheric Administration.

Kalkstein, L. S., P. F. Jamason, J. S. Greene, J. Libby, and L. Robinson. 1996. The Philadelphia Hot Weather-Health Watch/Warning System: Development and application. *Bulletin of the American Meteorological Society* 77:1519-1528.

Keetch, J. J., and G. Byram. 1968. A Drought Index for Forest Fire Control. Research Paper SE-38. Asheville, NC: U.S. Department of Agriculture, Forest Service, Southeastern Forest Experiment Station.

Le Houérou, H. N., G. F. Popov, and L. See. 1993. Agro-bioclimatic Classification of Africa. Agrometeorology Series Working Paper No. 6. Rome: Food and Agriculture Organization.

McKee, T. B., N. J. Doesken, and J. Kleist. 1993. The relationship of drought frequency and duration to time scales. *Preprints*, 8th Conference of Applied Climatology, Anaheim, CA, Jan. 17-22, pp. 179-194.

Palmer, W. C. 1965. Meteorological Drought. Research Paper No. 45. Washington, DC: U.S. Department of Commerce Weather Bureau.

Palmer, W. C. 1968. Keeping track of crop moisture conditions nationwide. The New Crop Moisture Index. *Weatherwise* 21:156-161.

Shafer, B. A., and L. E. Dezman. 1982. Development of a Surface Water Supply Index (SWSI) to assess the severity of drought conditions in snowpack runoff areas. Pp. 164-175 in *Proceedings of the Western Snow Conference*. Fort Collins, CO: Western Snow Conference.

Willeke, G., J. R. M. Hosking, J. R. Wallis, and N. B. Guttman. 1994. The National Drought Atlas. Institute for Water Resources Report No. 94-NDS-4. Alexandria, VA: U.S. Army Corps of Engineers.

APPENDIX B

Top States in Events and Losses by Individual Hazard

TABLE B-1 Flood Deaths and Damages, 1975-1998

State	No. of Deaths	State	Damages (millions, adjusted to U.S. $1999)
Texas	442	Texas	10,289.3
Pennsylvania	155	Louisiana	9,171.9
California	154	Iowa	9,074.1
Colorado	153	Missouri	6,355.2
Missouri	113	California	6,355.2
U.S. average	49.9	U.S. average	2,117.4

TABLE B-2 Tornado Events, Deaths, and Damages, 1975-1998

State	No. of Events	State	No. of Deaths	State	Damages (millions, adjusted to U.S. $1999)
Texas	3,510	Texas	246	Texas	3,645.7
Florida	1,498	Alabama	139	Arkansas	2,047.0
Oklahoma	1,195	Arkansas	110	Missouri	1,943.6
Kansas	1,119	Florida	78	Iowa	1,894.0
Nebraska	1,104	Mississippi	77	Florida	1,744.6
U.S. average	448.2	U.S. average	26.9	U.S. average	732.5

TABLE B-3 Hail Events and Damages, 1975-1998

State	No. of Events	State	Damages (millions, adjusted to U.S. $1999)
Texas	17,439	Texas	1,220.7
Oklahoma	12,572	Nebraska	901.8
Kansas	9,488	Kansas	405.8
Nebraska	5,151	Colorado	371.5
Missouri	4,166	Oklahoma	297.6
U.S. average	2,064.9	U.S. average	97.3

TABLE B-4 Thunderstorm Wind Events, Deaths, and Damages, 1975-1998

State	No. of Events	State	No. of Deaths	State	Damages (millions, adjusted to U.S. $1999)
Texas	10,844	Michigan	42	Texas	534.4
Oklahoma	6,644	Texas	38	Michigan	424.3
Ohio	5,525	Ohio	33	New York	340.3
Georgia	5,481	Louisiana	29	Iowa	249.8
Kansas	5,348	New York	29	Minnesota	212.9
U.S. average	2,533.3	U.S. average	9.4	U.S. average	80.1

TABLE B-5 Lightning Events and Damages, 1975-1998

State	No. of Fatalities	State	Damages (millions, adjusted to U.S. $1999)
Florida	205	Oregon	54.8
Texas	90	Texas	38.7
North Carolina	77	Wisconsin	29.5
Colorado	72	Pennsylvania	28.4
Ohio	66	New York	27.3
U.S. average	33.3	U.S. average	12.1

TABLE B-6 Wildfire Damages, 1975-1998

State	Damages (millions, adjusted to U.S. $1999)
California	826.8
Florida	387.4
New Mexico	113.7
Washington	96.2
Idaho	66.8
U.S. average	30.6

TABLE B-7 Extreme Heat Fatalities and Damages, 1975-1998

State	No. of Deaths	State	Damages (millions, adjusted to U.S. $1999)
Tennessee	153	Alabama	447.1
Alabama	125	Tennessee	184.3
Georgia	121	Georgia	176.6
Kentucky	42	Texas	102.0
Missouri	33	Ohio	89.7
U.S. average	11.3	U.S. average	21.0

TABLE B-8 Damage Totals for Drought Events, 1975-1998

State	Damages (millions, adjusted to U.S. $1999)
Iowa	3,379.0
Texas	2,680.0
Wisconsin	1,818.3
North Dakota	1,354.8
Arkansas	1,119.8
U.S. average	293.9

TABLE B-9 Winter Storm Deaths and Damages, 1975-1998

State	No. of Deaths	State	Damages (millions, adjusted to U.S. $1999)
New York	204	Alabama	5,805.7
California	75	Mississippi	5,648.7
Virginia	51	New York	2,203.1
Maryland	50	Iowa	592.5
South Carolina	43	Nebraska	522.0
U.S. average	21.0	U.S. average	398.6

TABLE B-10 Cold Weather Deaths and Damages, 1975-1998

State	No. of Deaths	State	Damages (millions, adjusted to U.S. $1999)
South Carolina	48	California	757.6
Ohio	30	Florida	565.3
North Carolina	22	Iowa	323.8
Louisiana	20	South Carolina	222.5
Indiana	18	Texas	187.3
U.S. average	4.6	U.S. average	57.0

TABLE B-11 Hurricane and Tropical Storm Events, Deaths, and Damages, 1975-1998[a]

State	No. of Events	State	No. of Deaths	State	Damages (millions, adjusted to U.S. $1999)
Florida	32	Florida	36	Florida	28,570.2
North Carolina	24	Texas	30	South Carolina	9,588.3
Georgia	21	North Carolina	25	Texas	3,634.9
South Carolina	17	Louisiana	14	Alabama	2,791.0
Texas	16	South Carolina	13	Mississippi	2,736.8
U.S. average	1.4	U.S. average	7.9	U.S. average	1,514.4

[a]Tropical storm (or hurricane) tracks as defined and maintained by the National Hurricane Center. One storm track may have affected multiple states. If a track crossed a state boundary, it was counted as an "event" for every state it crossed.

TABLE B-12 Earthquake Events, Deaths, and Damages, 1975-1998

State	No. of Events (epicenters)	State	No. of Deaths	State	Damages (millions, adjusted to U.S. $1999)
California	680,874	California	141	California	31,378.2
Nevada	23,809	Idaho	4	Idaho	49.0
Washington	21,982	Hawaii	2	Hawaii	23.5
Alaska	10,089	Oregon	2	Kentucky	2.0
Utah	8,839			Alaska	1.7

TABLE B-13 Selected Volcanoes, Eruption History, Deaths, and Damages

Volcano	No. of Eruptions During Past 200 Years	Last Eruption	No. of Deaths/ Injuries	Damages
Kilauea, Hawaii	47	1983-	80 (1790)	$12 million (1990)
Mauna Loa, Hawaii	30	1984		
Hualalai, Hawaii	1	1800-1801		
Mt. Baker, Washington	1	1870		
Mt. Rainier, Washington	1	1882		
Mt. St. Helens, Washington	2-3	1980-86	57	> $1 billion
Mt. Jefferson, Washington	0	50,000 BP		
Three Sisters, Oregon	0	950		
Crater Lake, Oregon	0	4,000 BP		
Mt. Shasta, California	1	1786		
Lassen Peak, California	1	1921		
Clear Lake, California	0	Unknown		
Long Valley Caldera, California	3	1850		
San Francisco Field, Arizona	2	1065-1180		
Bandera Field, New Mexico	1	1000		
Craters of the Moon, Idaho	1	2,100 BP		
Yellowstone Caldera, Wyoming, Montana, Idaho	0	70,000 BP		
Wrangell, Alaska	1	1902		
Redoubt Volcano, Alaska	4	Ongoing		$160 million (1989-1990)
Mt. Emmons-Pavlof Volcano, Alaska	40	1996-97		
Kiska Volcano, Alaska	7	1990		
Pyre Peak (Seguam), Alaska	5	1993		
Mt. Cleveland, Alaska	10	1994	1 (1944)	

Sources: FEMA 1997, *http://volcano.und.nodak.edu, http://vulcan.wr.usgs.gov, http://www.avo.alaska.edu, http://www.fema.gov.*

TABLE B-14 Hazardous Materials Accidents, Deaths, and Damages, 1975-1998

State	No. of Spills	State	No. of Deaths	State	Damages (millions, adjusted to U.S. $1999)
Pennsylvania	21,682	Florida	136	California	64.6
Ohio	19,577	Wisconsin	64	Texas	63.2
Illinois	16,810	Texas	46	Pennsylvania	45.8
California	16,184	Tennessee	36	Florida	31.8
Texas	15,883	California	23	Montana	29.0
U.S. average	5,187.7	U.S. average	11.6	U.S. average	15.5

TABLE B-15 Relative Hazardousness Associated with Nuclear Power Plants Based on SALP Reports, 1988-1998

State	No. of Facilities	SALP Score
Washington	1	8.58
New York	5	8.05
Illinois	7	7.66
Arizona	1	7.51
Michigan	4	7.46
Kansas	1	7.34
Maryland	1	7.28
Nebraska	2	7.27
North Carolina	2	7.12
New Jersey	3	7.06

TABLE B-16 Toxic Release Inventory Emissions, 1998

State	No. of TRI Facilities	State	Millions of Pounds Released
Ohio	1,682	Nevada	1,217.6
California	1,499	Arizona	1,068.6
Pennsylvania	1,389	Utah	573.0
Texas	1,381	Alaska	307.0
Illinois	1,344	Texas	288.0
U.S. average	466.2	U.S. total	7,300.0

TABLE B-17 Number of National Priority List Hazardous Waste Sites, 1998

State	No. of Sites
New Jersey	113
Pennsylvania	96
California	93
New York	86
Michigan	68
U.S. average	24.9

Index

C

D

E

F

G

H

I

T

Y